TRANSACTIONS

OF THE

AMERICAN PHILOSOPHICAL SOC

HELD AT PHILADELPHIA

FOR PROMOTING USEFUL KNOWLEDGE

VOLUME XXII—NEW SERIES

PART 2

The Secular Variations of the Elements of the Orbits of the Four Inner Planets (
Epoch 1850.0 G. M. T. By ERIC DOOLITTLE.

Philadelphia:

THE AMERICAN PHILOSOPHICAL SOCIET'

104 SOUTH FIFTH STREET

1912

ARTICLE II.

THE SECULAR VARIATIONS OF THE ELEMENTS OF THE ORBITS OF THE FOUR INNER PLANETS COMPUTED FOR THE EPOCH 1850.0 G. M. T.

By ERIC DOOLITTLE.

(Read March 1, 1912.)

TABLE OF CONTENTS.

37

To my Father,

PROFESSOR CHARLES L. DOOLITTLE,

this work

is Inscribed.

1. INTRODUCTION.

The usual method of determining the secular variations of the elements of any planet is the well-known one based upon the development of the perturbing function into an infinite series whose successive terms involve continually higher powers of the eccentricities and the mutual inclination. This method possesses two advantages. The first is that when an extreme degree of accuracy is not required, so that higher terms of the development may be disregarded, it is the simplest method available; and, in the second place, since the coefficients of all terms are general literal expressions, the change produced in the value of any variation by a change in the assumed values of one or more of the elements can readily be ascertained by a simple substitution of the more accurate values. On the other hand, this method possesses the disadvantage that the complexity of the expansion grows rapidly greater as the order of the included terms is increased, so that a slight increase in the desired accuracy greatly increases the labor of the computation.

The integral methods, founded upon the celebrated theorem of GAUSS[1],* are wholly free from this latter disadvantage, for if it is desired to include all terms to the twenty fourth order this can be done by a computation which is less than twice as long as that required when the approximation is stopped at terms of the eleventh order. But the integral method, though thus extremely accurate, leads only to the numerical values of the variations dependent upon the values of the elements assumed; if they are desired for some other epoch at which the various elements possess different values from those adopted, or if an improved value of any of the elements becomes known, they can only be found by an entire repetition of the computation.

The only determinations of the secular perturbations of the four inner planets which are in any sense modern ones are the classic investigation of LE VERRIER[7] and the computation of NEWCOMB[15]. The latter furnishes the most accurate values of these variations so far determined; the series were extended to terms of the eighth order, only those terms of this order being included, however, which seemed likely to be most important, and in some cases terms of the tenth order were included, though usually by induction merely.

In both of the above computations the usual expansion into an infinite series was employed. As the GAUSSIAN method is so extremely accurate, and as its formulas throughout are wholly different from those hitherto employed, it seemed that an

* These symbols wherever they occur refer to the list of titles at the end of the present paper.

application of it to a re-determination of these variations based upon the most accurate values of the several elements now obtainable would be of value. The results of this work will be found in the following pages; the final comparison with the earlier results is given in Article 11, and the comparison with the results of observation in Articles 12 and 13. The epoch throughout is 1850.0, G. M. T.

In the four following articles an attempt is made to state briefly the essential features of the various methods of computing secular variations which are founded on GAUSS's theorem, but for a detailed account of the long and often complex transformations which are involved, the original papers must be consulted.

2. THE METHOD OF GAUSS.

The equations which express the complete variations of the elements of the orbit of any body revolving about the sun when it is disturbed in its motion by the presence of a third body, may, as is well known, be put in a variety of different forms; the form selected as the basis for all developments founded on GAUSS's method[1] is that in which three rectangular components of the disturbing force enter into the expressions for the differential coefficients. Thus, if R denote the component lying in the direction of the radius vector of the disturbed body, positive outward from the sun; S, the component lying in the plane of the orbit of the disturbed body and perpendicular to the radius vector, positive in the direction of motion; and W, the component perpendicular to this plane and positive northward, we will have for the variation of the eccentricity of the orbit of the disturbed body,

$$\frac{de}{dt} = \frac{a^2 n \cos \varphi}{k^2(1+m)}[\sin v \cdot R + (\cos v + \cos E)S],$$

with similar expressions for the variations of the six remaining elements.*

In the original memoir of GAUSS the determination of the secular terms of these expressions was given a geometrical aspect. Thus, since each variation may obviously be expressed in terms of the two single variables M and M', the secular term in question will be that given by the equation,

$$\left[\frac{de}{dt}\right]_{00} = \frac{1}{4\pi^2}\int_0^{2\pi}\int_0^{2\pi}\frac{de}{dt}\,dM\,dM',$$

* The usual notation is adopted throughout. Thus a, $e = \sin \varphi$, i, Ω, π, n, and L are respectively the half major axis, the eccentricity, the inclination, the longitude of the ascending node, the longitude of perihelion, the mean motion and the longitude at the epoch of the disturbed body; M, E, v and r are respectively its mean, eccentric and true anomalies and its radius vector, m_0 is its mass, k^2 is the mass of the sun, and $m_0 = mk^2$. The same letters with accents refer to the disturbing body.

Watson, Theoretical Astronomy, pp. 516–523; Oppolzer, Lehrbuch zur Bahnbestimmung, Vol. II, p. 213; Tisserand, Mecanique Celeste, Vol. I, pp. 431–433, etc. The final forms of the equations expressing the other variations may be inferred from those stated at the end of Article 7.

and this is the same as,

$$\left[\frac{de}{dt}\right]_{00} = \frac{a^2 n \cos \varphi}{2\pi k^2 (1+m)} \int_0^{2\pi} \left[\sin v \cdot \frac{1}{2\pi} \int_0^{2\pi} R dM' + (\cos v + \cos E) \cdot \frac{1}{2\pi} \int_0^{2\pi} S dM'\right] dM,$$

since the variable of the first integration enters the expression only through R and S. In the equation as thus written R and S are supposed to contain the mass, m_0', as a factor so that if R_i and S_i are the corresponding values produced by a unit mass, $R = m_0' R_i$ and $S = m_0' S_i$.

If we now imagine an infinitely thin elliptic ring which coincides with the orbit of m', whose total mass is equal to the mass m_0', and the density of any portion of which is proportional to the time occupied by m' in describing that portion of its orbit, we will have for the three components of the attraction exerted by any portion dm_0',

$$R_i dm_0', \qquad S_i dm'_0, \qquad \text{and} \qquad W_i dm_0',$$

and integrating about the entire ring, we find for the complete components,

$$\int_0^{2\pi} R_i dm_0', \qquad \int_0^{2\pi} S_i dm_0', \qquad \text{and} \qquad \int_0^{2\pi} W_i dm_0'.$$

But by the conditions,

$$\frac{dm_0'}{m_0'} = \frac{dt}{T} = \frac{dM'}{2\pi},$$

and hence the components are,

$$\frac{1}{2\pi} \int_0^{2\pi} m_0' R_i dM', \qquad \frac{1}{2\pi} \int_0^{2\pi} m_0' S_i dM' \qquad \text{and} \qquad \frac{1}{2\pi} \int_0^{2\pi} m_0' W_i dM',$$

which are identical with

$$\frac{1}{2\pi} \int_0^{2\pi} R dM', \qquad \frac{1}{2\pi} \int_0^{2\pi} S dM' \qquad \text{and} \qquad \frac{1}{2\pi} \int_0^{2\pi} W dM'.$$

Thus the expressions giving the secular variations are seen to be the same whether these are derived from the moving planet or from the elliptic ring.*

The work of GAUSS contains no application to the determination of secular variations nor are all the formulas necessary for this purpose there developed; the first integration alone is effected, and it is shown that by changing first to the variable E' and afterward introducing a new variable, T, each of the complicated integrals may be made to depend upon elliptic integrals whose values GAUSS obtained by the introduction of a new algorithm called by him the Arithmetico-geometrical mean.

* Other interesting geometrical aspects of the problem are treated by Bour [5], Hill [37], [39], and Halphen [28], but for brevity a detailed account of these is here omitted,

The first application of Gauss's method was made by Nicolai[2], who determined by it the secular variations of the Earth's orbit, but the results only were published.* The first development of the method is by Clausen[3] who also applied it to a determination of the perturbations of Tuttle's Comet produced by the action of Jupiter[4], dividing the disturbed orbit into 120 parts with reference to the true anomaly. It was next, in 1867, applied by Adams[6] to the orbit of the November meteors with a special view to ascertaining the cause of the steady progression of the node of the orbit, but in this investigation certain small terms were neglected by Adams and the solution of a fundamental cubic equation which occurs in the original method was in this manner avoided.

No further applications of Gauss's method seem to have been made until after the publication of Hill's extensive development[8] and modifications of it in 1882.

3. HILL'S FIRST MODIFICATION OF GAUSS'S METHOD.

Although the first of the above integrations may be rigorously effected, the value of the second must be approximated to by a mechanical quadrature about the orbit of m, a greater or less number of terms being employed in the quadrature according as the disturbed orbit is more or less eccentric. Since either the true, eccentric, or mean anomalies may be selected as the variables, it becomes of importance to decide which of these must be chosen in order to render the quadrature most accurate. It is readily proved† that the inequalities of distribution of a series of points on an elliptic orbit corresponding to a series of equidistant values of the eccentric anomaly are of the order of the square of the eccentricity while for the other two anomalies they are of the order of the first power of this quantity, and therefore Hill has employed the eccentric anomalies throughout his development, although Seeliger[9] showed that a still higher accuracy will be obtained if the true anomalies are chosen. .

If, therefore, we decide to make the integrations with reference to the eccentric anomalies, we will obtain, since

$$dM = \frac{r}{a} dE, \qquad dM' = \frac{r'}{a'} dE', \qquad \text{and} \qquad r' = a'(1 - e' \cos E'),$$

$$\left[\frac{de}{dt}\right]_{\infty} = \frac{n \cos \varphi}{k^2(1 + m)} \cdot \frac{1}{4\pi^2} \int_0^{2\pi} \int_0^{2\pi} [\sin v \cdot Rar(1 - e' \cos E')$$
$$+ (\cos v + \cos E) \cdot Sar(1 - e' \cos E')]dEdE'dt,$$

---

* See Article 11. .

† See Tisserand's Mecanique Celeste, Vol. I, page 442.

and writing,

$$R_0 = \frac{1}{2\pi} \int_0^{2\pi} \frac{ar}{m_0'} R(1 - e' \cos E')dE', \qquad S_0 = \frac{1}{2\pi} \int_0^{2\pi} \frac{ar}{m_0'} S(1 - e' \cos E')dE',$$

$$W_0 = \frac{1}{2\pi} \int_0^{2\pi} \frac{ar}{m_0'} W(1 - e' \cos E')dE',$$

the expression for the secular variation will become,

$$\left[\frac{de}{dt}\right]_{00} = \frac{m'n}{1+m} \cos \varphi \cdot \frac{1}{2\pi} \int_0^{2\pi} [\sin v \cdot R_0 + (\cos v + \cos E)S_0]dE.$$

In order to find the values of R_0, S_0, and W_0, it is first necessary to express R, S and W in terms of E'. For this purpose that part of the disturbing force arising from the action of the disturbing planet upon the sun need not be included, for it is known that this has no secular term.† Considering therefore only the action of m' upon m, it is evident from a figure that R, S and W will have the values,

$$R = \frac{m_0'}{\Delta^2} \left\{ \frac{r' \cos \vartheta - r}{\Delta} \right\},$$

$$S = \frac{m_0'}{\Delta^2} \cdot \frac{r' \sin \vartheta \cos \gamma}{\Delta},$$

$$W = \frac{m_0'}{\Delta^2} \sin \gamma,$$

and also that

$$\Delta^2 = r^2 - 2rr' \cos \vartheta + r'^2,$$

in which ϑ is the angle included between the radii vectores, Δ is the distance between the two bodies, and γ is the inclination of the plane which includes r and r' to the plane of the orbit of the disturbed body.

If Π and Π' denote the angular distances respectively of the perihelia of the two orbits from the ascending node of the orbit of m' upon the orbit of m, and if I be their mutual inclination, we will have,

$$\cos \vartheta = \cos (v + \Pi) \cos (v' + \Pi') + \sin (v + \Pi) \sin (v' + \Pi') \cos I,$$

$$\sin \vartheta \cos \gamma = - \sin (v + \Pi) \cos (v' + \Pi') + \cos (v + \Pi) \sin (v' + \Pi') \cos I,$$

$$\sin \quad \sin \gamma = \sin I \sin (v' + \Pi')$$

The values of Π, Π', and I are obtained from the original elements by a direct solution of the spherical triangle whose sides are Π and Π', and in which the angle included between these sides is I. (See Article 7.)

† See Hill's "On Gauss's Method [(8)] . . . ," page 321.

If we now eliminate v' from the above expressions by the equations,

$$r' \cos v' = a' (\cos E' - e'), \qquad r' \sin v' = a' \cos \varphi' \sin E', \qquad r' = a' (1 - e' \cos E'),$$

the resulting equations giving R, S, W, and Δ will be expressed wholly in terms of the variable E'. In order to simplify these results, we assume certain new auxiliaries defined by the equations,

$$k \cos (K - \Pi) = \cos \Pi', \qquad k \sin (K - \Pi) = - \cos I \sin \Pi', \qquad k' \cos (K' - \Pi) = \cos I \cos \Pi',$$
$$k' \sin (K' - \Pi) = - \sin \Pi',$$
$$A = r^2 + 2ka'e'r \cos (v + K) + a'^2,$$
$$B \cos \epsilon = ka'r \cos (v + K) + a'^2 e',$$
$$B \sin \epsilon = k'a' \cos \varphi' \cdot r \sin (v + K'),$$

$$
\begin{aligned}
&A_c = ka' \cos (v + K), & &A_s = k'a' \cos \varphi' \sin (v + K') \\
&B_c = - ka' \sin (v + K), & &B_s = k'a' \cos \varphi' \cos (v + K') \\
&C_c = a' \sin \Pi' \sin I, & &C_s = a' \cos \varphi' \cos \Pi' \sin I. \\
& & &C = a'^2 e'^2,
\end{aligned}
$$

when the desired expressions become,

$$\frac{\Delta^3}{m_0'} R = A_c (\cos E' - e') + A_s \sin E' - r$$

$$\frac{\Delta^3}{m_0'} S = B_c (\cos E' - e') + B_s \sin E'$$

$$\frac{\Delta^3}{m_0'} W = C_c (\cos E' - e') + C_s \sin E'$$

$$\Delta^2 = A - 2B \cos (E' - \epsilon) + C \cos^2 E'.$$

In order to effect the integrations, Gauss here introduced a new variable, T, connected with E' by the relations,

$$N \sin E' = \alpha + \alpha' \sin T + \alpha'' \cos T$$
$$N \cos E' = \beta + \beta' \sin T + \beta'' \cos T$$
$$N = \gamma + \gamma' \sin T + \gamma'' \cos T,$$

the quantities α, α', α'', β, β' ... being subject to the conditions that

$$(N \sin E')^2 + (N \cos E')^2 - N^2 \qquad \text{and} \qquad \sin^2 T + \cos^2 T - 1$$

shall be identically zero, and also being so chosen that the coefficients of $\sin T$, $\cos T$, and $\sin T \cos T$ shall vanish in the expression $N^2 \Delta^2$ which therefore must take the form,

$$G - G' \sin^2 T + G'' \cos^2 T.$$

From these conditions it is derived that the coefficients G, G' and $- G''$ in the transformed expression for $N^2\Delta^2$ must severally satisfy the cubic equation,

$$x(x - A)(x + C) + B^2x + B^2C \sin^2 \epsilon = 0,$$

and hence that they must be the roots of this equation. By substituting for x the successive values, $- C$, 0, $a'^2 \cos^2 \varphi'$ and $+ A$, the first member is seen to take in succession the corresponding values,

$$- B^2C \cos^2 \epsilon$$
$$+ B^2C \sin^2 \epsilon$$
$$- a'^4 \cos^2 \varphi' \cdot r^2 \sin^2 I \sin^2 (v + \Pi)$$
$$+ B^2(A + C \sin^2 \epsilon).$$

Since, even when $\cos (v + K)$ has its maximum negative value, the value of A exceeds that of $(r - a')^2$, it is evident that A is always positive, and therefore that the above equation has one negative root which lies between $- C$ and 0, one positive root lying between 0 and $a'^2 \cos^2 \varphi'$, and that the third root lies between this value and $+ A$. The roots are represented by $- G''$, G', and G, respectively, and thus G'', G' and G are always positive quantities, the last being the largest and the first the smallest except when φ' exceeds $45°$, a case not met with in any of the planetary orbits.

Since α, β, γ, α', β' . . . must retain the same values whatever the values of E' and T, we may, by writing the equations arising from the three conditions above stated and equating the coefficients of the like terms in the two members, obtain a series of equations which are sufficient for the determination of these quantities in terms of G, G', G'' and the other known auxiliaries. Upon substituting the resulting expressions for $\sin E'$ and $\cos E'$ in the equations defining R_0, S_0, and W_0, and noticing that $N^2\Delta^2$ may be written,

$$G - G' \sin^2 T + G'' \cos^2 T = (G' + G'') \left\{ 1 - \frac{G' + G''}{G + G''} \sin^2 T \right\},$$

we obtain each of the components in the form,

$$R_0, \ S_0 \text{ or } W_0 = \frac{1}{2\pi} \int_0^{2\pi} \frac{m_s \sin^2 T + m_c \cos^2 T}{(G + G'')^{\frac{3}{2}} \left(1 - \frac{G' + G''}{G + G''} \sin^2 T \right)^{\frac{3}{2}}} dT.$$

If we now write,

$$\frac{G' + G''}{G + G''} = c^2 = \sin^2 \theta,$$

and consider that from LANDEN's well-known transformation,

$$F\left(c, \frac{\pi}{2} \right) = \frac{\pi}{2} K = \frac{\pi}{2} (1 + c_0)(1 + c_{00})(1 + c_{000}) \cdots$$

$$E\left(c,\frac{\pi}{2}\right) = \frac{\pi}{2}KL = \frac{\pi}{2}K\left(1 - \frac{c^2}{2} - \frac{c^2c_0}{4} - \frac{c^2c_0c_{00}}{8}\cdots\right),$$

and also notice that

$$\int_0^{\pi/2} \frac{dT}{(1 - c^2\sin^2 T)^{\frac{3}{2}}} = \sec^2\theta\cdot E\left(c,\frac{\pi}{2}\right)$$

$$\int_0^{\pi/2} \frac{\sin^2 T\,dT}{(1 - c^2\sin^2 T)^3} = \frac{1}{c^2}\left[\sec^2\theta\cdot E\left(c,\frac{\pi}{2}\right) - F\left(c,\frac{\pi}{2}\right)\right]$$

$$\int_0^{\pi/2} \frac{\cos^2 T\,dT}{(1 - c^2\sin^2 T)^{\frac{3}{2}}} = \frac{1}{c^2}\left[F\left(c,\frac{\pi}{2}\right) - E\left(c,\frac{\pi}{2}\right)\right],$$

it is evident that each of the above three integrals becomes expressible wholly in terms of the rapidly convergent series of LANDEN.

For the purposes of the present computation HILL[8] has computed to ten places the logarithms of the quantities

$$K_0 = \sec^2\theta\cdot KL, \qquad L'_0 = \frac{L - \cos^2\theta}{c^2L}, \qquad \text{and} \qquad N_0 = \sec^2\theta\cdot(1 + L_0'),$$

and these correct to eight places are tabulated at intervals of one tenth of a degree for all values of θ from $\theta = 0°$ to $\theta = 50°$.

From a direct substitution it is now seen that the final resulting values of R_0, S_0 and W_0 are as follows, in which the symbols N, P, Q, etc., are written for abbreviation and have the meanings stated in Article 7:

$$R_0 = -N - QG' + VJ_1',$$
$$S_0 = PF_2 + VJ_2$$
$$W_0 = PF_3 + VJ_3$$

The integration with respect to E' having been thus entirely completed, that in regard to E is effected by mechanical quadratures. Since each variation is a function of E alone, it follows by the principles of quadratures that if any one of them be expanded into a periodic series involving the sines and cosines of E and its multiples, the secular term of the series, which is rigorously equal to $\frac{1}{2}\pi\int_0^{2\pi} f(E)dE$, may be also obtained by forming the values of $f(E)$ for $2j$ equidistant values of E, from $E = 0°$ to $E = 360°$, and dividing the sum by $2j$. The expression thus obtained,

$$\frac{1}{2j}\cdot\Sigma f(E),$$

will be subject only to the error involved in dropping those terms which contain a multiple of E not lower than $2j$. An inspection of the known forms of the series which express the variations renders it evident that the error thus committed is of the order $2j$ in terms of the eccentricities and mutual inclinations of the orbits except in the one case of the variation of the Mean Longitude, in which, as this variation depends wholly upon the expansion of $-2(r/a)R_0$, it is of the order $2j + 1$.

The resulting equations giving the values of all the secular variations are those stated in Article 7.

4. HILL'S SECOND MODIFICATION OF GAUSS'S METHOD. THE WORK OF CALLANDREAU AND INNES.

In HILL's second modification of GAUSS's method[8], the well-known expressions for the roots of a cubic equation when this is solved by the trigonometric method are introduced, and thus, throughout the integrals, the quantities p, q and θ' occur instead of the roots G, G' and G'', the equations connecting these quantities being,

$$G = 2q \sin\left(60° - \frac{\theta'}{3}\right) + p, \qquad G' = 2q \sin\frac{\theta'}{3} + p,$$
$$G'' = 2q \sin\left(60° + \frac{\theta'}{3}\right) - p.$$

It was shown in GAUSS's original memoir[1] that

$$\int_0^{\pi/2} \frac{dT}{(m^2 \cos^2 T + n^2 \sin^2 T)^{\frac{1}{2}}} = \int_0^{\pi/2} \frac{dT}{(m'^2 \cos^2 T + n'^2 \sin^2 T)^{\frac{1}{2}}},$$

if $m' = \frac{1}{2}(m + n)$ and $n' = \sqrt{mn}$, and that by repeating this transformation by the employment of the equations,

$$m'' = \tfrac{1}{2}(m' + n'), \qquad n'' = \sqrt{m'n'},$$
$$m''' = \tfrac{1}{2}(m'' + n''), \qquad n''' = \sqrt{m''n''},$$
$$\text{etc.} \qquad\qquad \text{etc.,}$$

$m^{(k)}$ and $n^{(k)}$ very rapidly approach a single limit, μ, which GAUSS named the Arithmetico-geometrical Mean. It thus follows that our first integral is equal to $\pi/2\mu$, and that integrals of the form

$$\int_0^{\frac{\pi}{2}} \frac{(\sin^2 T - \cos^2 T)dT}{(m^2 \cos^2 T + n^2 \sin^2 T)^{\frac{1}{2}}}$$

become equal to $\pi/2 \cdot \omega/\mu$ in which ω is a very rapidly converging series involving m, n, m', n', etc., in its successive terms.

The integral expressions which actually enter into the equations for R_0, S_0, and W_0 are

$$\chi(\theta') = \frac{\sin\left(60° - \frac{\theta'}{3}\right) - \frac{\sqrt{3}}{4} \cdot \frac{\omega}{(m^2 - n^2)}}{24 \sqrt[4]{12}\, m^4 n^4 \mu},$$

$$\psi(\theta') = \frac{\frac{1}{2} + m^2n^2 - \dfrac{\sqrt{3}}{4} \cdot \dfrac{\omega \sin \theta'}{(m^2 - n^2)}}{24 \sqrt[4]{12}\, m^4 n^4 \mu},$$

in which

$$m^2 = \cos \frac{\theta'}{3}$$

and

$$n^2 = \cos \left(60° + \frac{\theta'}{3}\right), \quad \cdot$$

the values of R_0, S_0, and W_0 being connected by comparatively simple relations with these quantities and with known auxiliaries.

HILL accordingly suggested that tables of these functions should be computed, and this was first done by MONS. O. CALLANDREAU[13] who however adopted as an argument the quantity α defined by the relation

$$\frac{1 - \alpha}{1 + \alpha} = \sqrt{\frac{\cos \left(60° + \dfrac{\theta'}{3}\right)}{\cos \dfrac{\theta'}{3}}}$$

and tabulated the logarithms of the functions $m^4 n^4 \chi(\theta')$ and $\psi(\theta') \div \chi(\theta')$ at intervals of 0.001 from $\alpha = 0.000$ to $\alpha = 0.400$; of 0.002 from $\alpha = 0.400$ to $\alpha = 0.600$ and of 0.005 from this point to the extreme value, $\alpha = 1.000$. This paper repeats the derivation of all formulas necessary when the second method alone is employed, essentially as this was given by HILL, and also contains a direct proof that R_0, S_0, and W_0 can be expressed wholly in terms of the complete elliptic integrals, F and E.

Similar tables were also computed by MR. R. T. A. INNES[22], the functions here tabulated being $(1 - \alpha)/(1 + \alpha^4) \cdot \psi(\theta')$ and $\psi(\theta') \div \chi(\theta')$ to the argument θ' at intervals of one degree, from $\theta' = -90°$ to $\theta' = +90°$.

Whether the first or second methods be employed, the values of the integrals involved may also, as was pointed out by HILL[38], be approximated to with great rapidity by the use of JACOBI's Nome, q (American Journal of Mathematics, Vol. 23, page 321. In the Astronomical Journal, No. 511, a brief application is given to a case in the action of Venus on the Earth). This function is defined by the equation,

$$q = e^{-\pi \frac{K'}{K}},$$

in which K' is the complete elliptic integral of the first kind complementary to K, from which there may be derived,

$$K = \frac{\pi}{2}(1 + 2q + 2q^4 + 2q^9 + 2q^{16} \cdots)^2$$

$$KE = (1 + 3^2 q^2 + 5^2 q^6 \cdots) \div (1 + q^2 + q^6 \cdots).$$

The values of log $[q/\tan^2 \frac{1}{2}\theta]$ computed to ten decimal places for each degree of θ from $\theta = 0°$ to $\theta = 45°$ are given by INNES[39]. When θ exceeds $45°$, the values of K and E are readily obtained from their expressions in terms of the complementary complete integrals whose moduli are $\sin(\pi/2 - \theta)$, and to which the table is therefore directly applicable.

Lastly, in the second method, HILL recommends that the quadratures be performed upon the quantities $a/r \cdot R_0$, $a/r \cdot S_0$ and $a/r \cdot W_0$ directly, all constant and evanescent factors which appear in the expressions for the variations being removed from under the integral signs and reserved until the integration has been completed.

5. THE METHOD OF HALPHEN AND ITS MODIFICATIONS BY ARNDT AND INNES.

It was first pointed out by BRUNS[29] that the periods of the elliptic functions of the first and second integrals can be evaluated without a knowledge of the three roots, but it was HALPHEN[28] who first applied this remarkably elegant method of analysis to the present problem. It was shown by him that if ω and η are the two periods in question, then R_0, S_0, and W_0 may be obtained in the form $a\omega + b\eta$, in which a and b are rational functions of the coefficients of the cubic equation and ω and η are expressible in terms of certain hyper-geometric series in which the common variable is an absolute invariant of the elliptic functions.

The three integrals entering into the problem have the form,

$$\int_0^{\pi/2} \frac{l\,dT}{(G - G' \sin^2 T + G'' \cos^2 T)^{\frac{1}{2}}},$$

in which l has the values 1, $\sin^2 T$ and $\cos^2 T$, respectively, in the three cases; by introducing the new variable, s, defined by the relation,

$$\sin^2 T = \frac{G + G''}{G' + G''} \cdot \frac{s - G'}{s - G},$$

these become,

$$-2\frac{G' + G''}{n} \int_{G'}^{-G''} \frac{ds}{\sqrt{S}}(s - G), \qquad -2\frac{G + G''}{n} \int_{G'}^{-G''} \frac{ds}{\sqrt{S}}(s - G')$$

and

$$-2\frac{G' - G}{n} \int_{G'}^{-G''} \frac{ds}{\sqrt{S}}(s + G''),$$

respectively, in which

$$n = (G + G'')(G - G')(G' + G'')$$

and

$$S = -4(s - G)(s - G')(s + G'').$$

Introducing the WEIERSTRASSIAN $\wp$ function through the relation

$$u = \int_{\wp(u)}^{s} \frac{ds}{\sqrt{4(s - e_1)(s - e_2)(s - e_3)}},$$

u being the elliptic integral of the first kind and e_1, e_2, and e_3 the roots of the cubic equation increased by one third of the coefficient of x^2, and considering that from the theory of these functions,

$$s - G = \wp(u) - e_1, \quad \wp(\omega) = e_1, \quad \wp(\omega + \omega') = e_2, \quad \text{and} \quad \wp(\omega') = e_3,$$

the first integral will become,

$$2\frac{G' + G''}{n} \int_{\omega}^{\omega+\omega'} [\wp(u) - e_1]du = 2\left[e_1\omega + \frac{\sigma'}{\sigma}(\omega + \omega') - \frac{\sigma'}{\sigma} \cdot \omega' \right],$$

σ and σ' being the second WEIERSTRASSIAN functions, which are connected with the periods, ω and η, by the equations,

$$\frac{\sigma'}{\sigma}(\omega + \omega') = \eta + \eta'; \quad \frac{\sigma'}{\sigma}\omega' = \eta'.$$

The three integrals consequently take the final forms,

$$2\frac{G' + G''}{n}(e_1\omega + \eta); \quad 2\frac{G + G''}{n}(e_2\omega + \eta), \quad \text{and} \quad 2\frac{G' - G}{n}(e_3\omega + \eta).$$

A direct substitution of these expressions for the integrals in the equations which define R_0, S_0 and W_0 leads, after some reduction, to forms which are seen to contain only these integrals themselves, the coefficients of the cubic equation with other known auxiliaries, and the quantity n. But if, for brevity, we write the original cubic equation in the form,

$$x^3 - P_1 x^2 + P_2 x - P_3 = 0,$$

and let

$$\lambda = P_1^2 - 3P_2 \quad \text{and} \quad \rho = P_1 P_2 - 9P_3,$$

then the invariants, g_2 and g_3, and the absolute invariant, g, will have the values,

$$g_2 = \tfrac{1}{3}\lambda; \quad g_3 = \tfrac{1}{27}(2P_1\lambda - 3\rho), \quad \text{and} \quad g = g_2^3 \div 27g_3^2,$$

and n will be given by,

$$n^2 = \tfrac{1}{16}(g_2^3 - 27g_3^2),$$

in which the last factor is the discriminant. Thus, except for ω and η, our final expressions are obtained wholly in terms of the coefficients of the cubic equation, and a knowledge of the roots becomes unnecessary.

In the paper by BRUNS, before referred to, it is shown that ω and η are directly expressible in terms of a hyper-geometric series whose variable is the absolute invariant, g. By a simple transformation the relations may be placed in the following forms, which are more convenient in practical application.

$$\omega = \frac{\pi}{\sqrt[4]{12}} g_2^{-\frac{1}{4}} F\left(\frac{1}{12}, \frac{5}{12}, 1, \frac{g-1}{g}\right),$$

$$\eta = \frac{\pi}{\sqrt[4]{1728}} g_2^{\frac{1}{4}} F\left(\frac{1}{12}, -\frac{1}{12}, 1, \frac{g-1}{g}\right).$$

Dr. Louis Arndt[30] has fully developed this method, deriving all the formulas necessary for its application and stating tables for $F(\omega)$ and $F(\eta)$ for values of $(g-1)/g$ from $(g-1)/g = 0.000$ to $(g-1)/g = 0.980$, the interval being 0.001.

In a recent paper by Innes[31] the complete formulas for this method are derived when the quadrature is applied directly to the expressions $(a/r)R_0$, $(a/r)S_0$ and $(a/r)W_0$, as suggested in the second method of Hill. The development is nearly identical with that of Arndt except that the forms of the hyper-geometric series are slightly changed, the variable,

$$\sin^2\frac{i}{2} = \frac{\sqrt{g}-1}{2\sqrt{g}}$$

being preferred. The values of the logarithms of

$$F(\omega) = F\left(\frac{1}{6}, \frac{5}{6}, 2, \sin^2\frac{i}{2}\right) \quad \text{and} \quad F(\eta) = F\left(-\frac{1}{6}, \frac{5}{6}, 2, \sin^2\frac{i}{2}\right).$$

were published by Mr. Frank Robbins[32] for all values of i, at intervals of one degree from $i = 1°$ to $i = 90°$, the computation having been made to ten places and published to seven, and these tables, computed with seven place logarithms, have been extended from $i = 90°$ to $i = 180°$ by Mr. C. J. Merfield[33].

Although the preceding methods are of great mathematical elegance, it is doubtful whether their formulas lead to so accurate results as those of Hill's first method when seven place logarithms are employed. (See the computations of Jupiter on Mars[24] and of Saturn on Mars[25], Article 10.) Moreover, when the method is applied which is explained in the computation of Jupiter on Mercury (Article 10), the roots of the cubic equation are so readily obtained that the avoidance of its solution becomes a matter of no practical importance. Accordingly Hill's first modification of Gauss's method has been employed throughout all of the following computation.

THE COMPUTATION.

6. THE ELEMENTS OF THE ORBITS AND THE ADOPTED MASSES.

The values adopted for the elements of the several orbits, to serve as the basis for this computation, were taken in each case from HILL's "*New Theory of Jupiter and Saturn.*"[16] Those of the four inner planets will be found on page 192; those of Jupiter and Saturn on page 558; of Uranus on page 109, and of Neptune on page 161. The epoch throughout is 1850.0 G. M. T.

The values of the masses finally selected by HILL, and here adopted, will be found on page 554 for Mercury, Venus and the Earth; on page 192 for Mars; on page 19 for Jupiter and Saturn, and on page 161 for Neptune. The mass of Uranus as stated in the "*New Theory*" is $1 \div 22640$, but at DR. HILL's suggestion this is here diminished to $1 \div 22800$, (*A. J.*, No. 316). The value assumed for the mass of Mercury when the first of these computations were made was $1 \div 5000000$, but all of the results are here changed to agree with the value $1 \div 7500000$ stated below. It seems not improbable that even this latter fraction is too large, but the true value of this element is still very uncertain.

	π			i			Ω			e	n
Mercury	75°	7′	13.62″	7°	0′	7.71″	46°	33′	8.63″	0.20560476	5381016.260″
Venus	129	27	42.83	3	23	35.01	75	19	53.08	0.00684311	2106641.357
Earth	100	21	39.73	0	0	0.00		—		0.01677114	1295977.416
Mars	333	17	51.74	1	51	2.24	48	23	54.59	0.09326803	689050.784
Jupiter	11	54	31.67	1	18	42.10	98	56	19.79	0.04825511	109256.626
Saturn	90	6	41.37	2	29	40.19	112	20	49.05	0.05606025	43996.21506
Uranus	168	15	6.70	0	46	20.54	73	14	8.00	0.0469236	15425.752
Neptune	43	17	30.30	1	47	1.68	130	7	31.83	0.0084962	7864.935

	$\log a$	$1 \div m$
Mercury	9.5878217	7 500 000
Venus	9.8593378	408 134
Earth	0.0000000	327 000
Mars	0.1828971	3 093 500
Jupiter	0.7162374	1 047.879
Saturn	0.9794956	3 501.6
Uranus	1.2831044	22 800
Neptune	1.4781414	19 700

7. THE FORMULAS EMPLOYED IN THE COMPUTATION.

The following formulas are written in the order in which they were applied. When the right hand member appears in two different forms, one of these was used in the first computation and the other in the duplication, though sometimes other obvious modifications were made use of in the several cases differing from those which are here written.

The values of I, Π, and Π' were obtained from the general equations:

$$\sin I \sin (\Pi - \omega) = - \sin i' \sin (\Omega' - \Omega),$$
$$\sin I \cos (\Pi - \omega) = - \sin i \cos i' + \cos i \sin i' \cos (\Omega' - \Omega)$$
$$= \cos i \cos i' [- \tan i + \tan i' \cos (\Omega' - \Omega)],$$
$$\sin I \sin (\Pi' - \omega') = - \sin i \sin (\Omega' - \Omega),$$
$$\sin I \cos (\Pi' - \omega') = \cos i \sin i' - \sin i \cos i' \cos (\Omega' - \Omega),$$
$$= \cos i \cos i' [\tan i_{\prime} - \tan i \cos (\Omega' - \Omega)].$$

When the Earth is the disturbing body, these become,

$$I = i; \qquad \Pi = 180° + \omega; \qquad \Pi' = 180° + \pi' - \Omega;$$

and when the Earth is the disturbed body,

$$I = i'; \qquad \Pi = \pi - \Omega'; \qquad \Pi' = \pi' - \Omega'.$$

As i, i' and I are always small, eight place logarithms were generally here used to insure the accuracy of Π and Π'.

The auxiliaries k, k', K, K' and C were then found from the relations:

$$k \sin (K - \Pi) = - \cos I \sin \Pi'; \qquad k \cos (K - \Pi) = \cos \Pi'; \qquad k' \sin (K' - \Pi) = - \sin \Pi';$$
$$k' \cos (K' - \Pi) = \cos I \cos \Pi'; \qquad C = a'^2 e'^2,$$

and their values were tested by the equations,

$$\tan \sigma = \frac{k}{k'}; \qquad \tan \tfrac{1}{2}(K - K' + 90°) \cot \tfrac{1}{2}(K + K' - 90° - 2\Pi) = \frac{\sin (\Pi' - \sigma)}{\sin (\Pi' + \sigma)},$$
$$\sin (K - K') = \sin I \tan I \sin (K' - \Pi) \sin (K - \Pi) \cot \Pi'.$$

The orbit of the disturbed planet being then divided into $2j$ parts in regard to the eccentric anomaly, the following equations were applied to each point of division, of which those marked with an asterisk are test equations upon the sums of the functions corresponding respectively to the odd and even points of division of the orbit. The sums corresponding to the odd points are designated by Σ_1, those to the even points by Σ_2, and

$$\Sigma = \Sigma_1 + \Sigma_2.$$

$$r \sin v = a \cos \varphi \sin E,$$
$$r \cos v = a (\cos E - e),$$
$$r^2 = a^2 (1 - 2 e \cos E + e^2 \cos^2 E),$$

(the last equation giving the value of r^2 for use in A, N, and J_3. Since

$$\tfrac{1}{2} \log r^2 = \log r,$$

this affords also an independent test of r).

$$*\Sigma_1 v + 180° = \Sigma_2 v; \qquad *\Sigma_1 r = \Sigma_2 r = ja.$$

$$A = r^2 + 2ka'e'r \cos (v + K) + a'^2 = [r + ka'e' \cos (v + K)]^2 + a'^2[1 - k^2e'^2 \cos^2 (v + K)],$$

(the second form used with ZECH's tables in the duplication).

$$*\Sigma_1 A = \Sigma_2 A = ja^2 + \tfrac{1}{2}ja^2e^2 + j[a'^2 - 2kaa'ee' \cos K]$$
$$B \sin \epsilon = k'a' \cos \varphi'r \sin (v + K')$$
$$B \cos \epsilon = ka'r \cos (v + K) + a'^2e'$$
$$*\Sigma_1 B \sin \epsilon = \Sigma_2 B \sin \epsilon = - jk'aa' \cos \varphi' \cdot e \sin K'$$
$$*\Sigma_1 B \cos \epsilon = \Sigma_2 B \cos \epsilon = j[a'^2e' - kaa'e \cos K]$$
$$g = B^2C \sin^2 \epsilon$$

To effect the solution of the cubic equation, h and l were found from the equations,

$$h = \tfrac{1}{2}[A - C + \sqrt{(A + C)^2 - 4B^2}]; \qquad l = \tfrac{1}{2}[A - C - \sqrt{(A + C)^2 - 4B^2}],$$

the very convenient test equation, $hl = B^2 - AC$, being applied to each pair of values. The first approximation to G was then obtained from

$$G = h - \frac{g}{h(h - l)},$$

and further approximations by successive applications of

$$G = h - \frac{g}{G(G - l)}.$$

(The number of trials required never exceeded three.) G' and G'' then follow from the equations,

$$G' = \frac{1}{2}\left[(A - C - G) + \sqrt{(A - C - G)^2 + \frac{4g}{G}} \right]; \qquad G'' = \frac{g}{GG'},$$

and we have for verification,

$$G + G' - G'' = A - C; \qquad G' = h + \frac{g}{G'(h - G')}; \qquad G'' = \frac{g}{(h + G'')(l + G'')};$$
$$*\Sigma G + \Sigma G' - \Sigma G'' = \Sigma A - 2jC.$$

(In some cases the first approximations to G were found by,

$$p = \tfrac{1}{3}(A - C); \quad q^2 = p^2 - \tfrac{1}{3}(B^2 - AC); \quad r = \tfrac{1}{2}p(p^2 - 3q^2) + \tfrac{1}{2}g;$$

$$\sin \theta' = \frac{r}{q^{\frac{3}{2}}}; \quad G = 2q \sin (60° - \tfrac{1}{3}\theta') + p,$$

the solution being then finished as before).

The modulus, $(c = \sin \theta)$, of the elliptic integrals employed in the computation was separately found by the two equations,

$$\sin^2 \theta = \frac{G' + G''}{G + G''}; \quad \tan^2 \theta = \frac{G' + G''}{G - G'},$$

and with θ as an argument the values of $\log K_0$, $\log L_0'$, and $\log N_0$ were taken from the tables of HILL's memoir[3], the interpolation being effected in both directions to second differences by the well-known formulas,

$$f(\theta_\kappa + nw) = f(\theta_\kappa) + n\left(D'_{\kappa+\frac{1}{2}} - \frac{1 - n}{2} D''_\kappa\right),$$

$$f(\theta_{\kappa+1} - n'w) = f(\theta_{\kappa+1}) - n'\left(D'_{\kappa+\frac{1}{2}} + \frac{1 - n'}{2} D''_{\kappa+1}\right),$$

in which $n + n' = 1$.

The logarithms of N, P, Q, and V were then obtained from,

$$N = \frac{ar^2 K_0}{(G + G'')^{\frac{1}{2}}}; \quad P = \frac{NL_0'}{(G + G'')^2}; \quad Q = \frac{NN_0}{G + G''}; \quad V = Q - PG'',$$

the first three being verified by similar operations performed upon the values of Σ_1 and Σ_2 formed from the respective logarithms, and the last by the use of ZECH's tables and also by the equation,

$$V = ar^2(G + G'')^{-\frac{1}{2}}[GN_0 + G''(N_0 - L_0')]K_0.$$

The following auxiliaries were next obtained:

$$J_1' = a'^2 \cos^2 \varphi'[1 - \sin^2 I \sin^2 (v + \Pi)] + G''$$
$$= [a' \cos \varphi' + a' \cos \varphi' \sin I \sin (v + \Pi)][a' \cos \varphi' - a' \cos \varphi' \sin I \sin (v + \Pi)] + G'',$$
$$J_2 = ka'e'r \sin (v + K) - \tfrac{1}{2}a'^2 \cos^2 \varphi' \sin^2 I \sin 2(v + \Pi)$$
$$= ka'e'r \sin (v + K) - a'^2 \cos^2 \varphi' \sin (v + \Pi) \cos (v + \Pi) \sin^2 I,$$

the second form being employed with ZECH's tables in the duplication

$$J_3 = \frac{a'^2}{a} \cos^2 \varphi' \sin I \cos I \cdot r \sin (v + \Pi) - \frac{a'}{a} e' \sin I \sin \Pi' \cdot r^2,$$

$$*\Sigma_1 J_3 = \Sigma_2 J_3 = - ja^2 \cos^2 \varphi' \sin I \cos I \cdot e \sin \Pi - \frac{a'e'}{a} \sin I \sin \Pi' \cdot \Sigma_1 r^2.$$

$$F_2 = - a'^2 \sin \varphi' \cos \varphi' \cos I \cdot B \sin \epsilon,$$

$$*\Sigma_1 F_2 = \Sigma_2 F_2 = jk'aa'^3 ee' \cos^2 \varphi' \cos I \sin K',$$

$$F_3 = -\frac{a'^2}{a} \sin \varphi' \cos \varphi' \sin I \cdot r \cos (v + \Pi) \cdot B \sin \epsilon.$$

There were next obtained,

$$R_0 = -N - QG' + VJ_1'; \qquad S_0 = PF_2 + VJ_2; \qquad W_0 = PF_3 + VJ_3;$$

$$R^{(n)} = \frac{1}{r} R_0 \sin E; \qquad S^{(n)} = \frac{1}{r} S_0; \qquad \tfrac{1}{2} A_1^{(s)} = \frac{a}{2j} \Sigma R^{(n)}; \qquad B_0^{(c)} = \frac{a}{2j} \Sigma S^{(n)},$$

and the very accurate test equation,

$$\sin \varphi \cdot \tfrac{1}{2} A_1^{(s)} + \cos \varphi \cdot B_0^{(c)} = 0,$$

was applied.

These values were then substituted in the following series of equations, and the final values of the differential coefficients obtained:

$$\left[\frac{de}{dt}\right]_{00} = \frac{m'n}{1 + m} \cdot \cos \varphi \cdot \frac{1}{2j} \Sigma[\sin v \cdot R_0 + (\cos v + \cos E)S_0],$$

$$\left[\frac{d\chi}{dt}\right]_{00} = \frac{m'n}{1 + m} \cdot \frac{\cos \varphi}{e} \cdot \frac{1}{2j} \Sigma\left[- \cos v \cdot R_0 + \left(\frac{r}{a} \sec^2 \varphi + 1\right) \sin v \cdot S_0 \right],$$

$$\left[\frac{di}{dt}\right]_{00} = \frac{m'n}{1 + m} \cdot \sec \varphi \cdot \frac{1}{2j} \Sigma[\cos u \cdot W_0],$$

$$\left[\frac{d\Omega}{dt}\right]_{00} = \frac{m'n}{1 + m} \cdot \frac{\sec \varphi}{\sin i} \cdot \frac{1}{2j} \Sigma[\sin u \cdot W_0],$$

$$\left[\frac{d\pi}{dt}\right]_{00} = \left[\frac{d\chi}{dt}\right]_{00} + 2 \sin^2\frac{i}{2} \left[\frac{d\Omega}{dt}\right]_{00},$$

$$\left[\frac{dL}{dt}\right]_{00} = \frac{m'n}{1 + m} \cdot \frac{1}{2j} \Sigma\left[- 2\frac{r}{a} R_0 \right] + 2 \sin^2\frac{\varphi}{2} \left[\frac{d\chi}{dt}\right]_{00} + 2 \sin^2\frac{i}{2} \left[\frac{d\Omega}{dt}\right]_{00}.$$

When the Earth is the disturbed body, the third and fourth equations are replaced by,

$$\left[\frac{dp}{dt}\right]_{00} = \frac{m'n}{1 + m} \cdot \sec \varphi \cdot \frac{1}{2j} \Sigma[\sin (v + \pi) \cdot W_0]$$

$$\left[\frac{dq}{dt}\right]_{00} = \frac{m'n}{1 + m} \cdot \sec \varphi \cdot \frac{1}{2j} \Sigma[\cos (v + \pi) \cdot W_0]$$

In this case

$$\left[\frac{d\chi}{dt}\right]_{00} = \left[\frac{d\pi}{dt}\right]_{00},$$

and the last term of the expression for $[dL/dt]_{00}$ disappears, but the first two equations remain unaltered.

8. THE VALUES OF THE PRELIMINARY CONSTANTS.

The values obtained for those constants which are direct functions of the elements of the orbits in the several cases are shown in the following tables. The last columns of these tables contain the differences between the values of $K - K'$ formed directly and the same angles obtained from the test formula of the preceding article. The other test equations were also exactly satisfied.

An examination of the formulas of the preceding article renders it evident that with any two planets I will have the same value whether the inner or the outer planet is the disturbing one, while the value of Π in the first case will differ 180° from that of Π' in the second, and that of Π' in the first case will similarly differ 180° from that of Π in the second. These conditions will be seen to be here satisfied very exactly, the minute discrepancies which occur being due to the fact that in some places eight place logarithms were employed, in others seven, and in still others the attainment of a higher accuracy throughout the entire computation was sought by the use of the dash, (—), which was placed above the last figure of each logarithm for which the interpolation led to a value coinciding more nearly with the mean of the two adjacent figures than with either one of them. In combining such logarithms the effect of the dash was taken into consideration by methods which are obvious.

Mercury by—	I			Π			Π'			K		
	°	′	″	°	′	″	°	′	″	°	′	″
Venus	4	20	42.982	230	39	31.39	284	54	1.27	305	43	2.40
Earth	7	0	7.710	208	34	4.99	233	48	31.10	334	57	50.59
Mars	5	9	10.165	209	13	54.31	107	24	19.31	101	45	36.16
Jupiter	6	17	15.310	218	5	54.72	154	49	24.01	63	8	32.52
Saturn	6	23	44.130	229	26	43.69	244	17	50.53	345	17	17.36
Uranus	6	19	17.399	211	43	10.39	304	49	47.06	266	43	32.95
Neptune	7	1	42.654	223	12	39.15	191	16	25.42	324	34	3.91

Mercury by—	K'			log k	log k'	log C	resid.
	°	′	″				″
Venus	305	47	57.49	9.9988328	9.9999176	5.3891826	0.007
Earth	334	33	18.85	9.9978879	9.9988719	6.4491252	0.002
Mars	101	53	33.05	9.9983990	9.9998432	8.3052599	0.004
Jupiter	63	24	30.76	9.9995281	9.9978563	8.7995614	0.002
Saturn	345	0	30.90	9.9978013	9.9994926	9.4563012	0.013
Uranus	267	3	12.48	9.9982188	9.9991396	9.9089914	0.000
Neptune	324	24	7.31	9.9968502	9.9998757	8.8147322	0.001

Venus by—	I			Π			Π'			K		
	°	′	″	°	′	″	°	′	″	°	′	″
Mercury	4	20	42.980	104	54	1.27	50	39	31.37	54	19	21.08
Earth	3	23	35.010	234	7	49.75	205	1	46.65	29	8	21.75
Mars	1	56	2.460	208	26	43.81	52	18	22.07	156	9	18.63
Jupiter	2	15	11.352	247	36	52.56	130	2	45.43	117	32	48.56
Saturn	2	3	12.046	281	7	33.71	241	43	52.16	39	24	36.81
Uranus	2	37	16.883	233	30	46.37	272	18	13.25	321	12	24.44
Neptune	2	46	38.369	265	47	34.23	179	34	46.34	86	12	46.11

Venus by—	K′			log k	log k′	log C	resid.
	°	′	″				″
Mercury	54	9	38.85	9.9992531	9.9994984	7.8017097	0.000
Earth	29	3	44.28	9.9998637	9.9993746	6.4491252	0.003
Mars	156	7	24.89	9.9998450	9.9999075	8.3052599	0.000
Jupiter	117	35	25.70	9.9998033	9.9998610	8.7995614	0.001
Saturn	39	22	46.29	9.9997836	9.9999375	9.4563012	0.003
Uranus	321	12	41.79	9.9995460	9.9999992	9.9089914	0.001
Neptune	86	12	49.67	9.9999999	9.9994896	8.8147322	0.001

Earth by—	I			П			П′			K		
	°	′	″	°	′	″	°	′	″	°	′	″
Mercury	7	0	7.710	53	48	31.10	28	34	4.99	25	25	13.33
Venus	3	23	35.010	25	1	46.65	54	7	49.75	330	56	48.79
Mars	1	51	2.240	51	57	45.14	284	53	57.15	127	3	21.25
Jupiter	1	18	42.100	1	25	19.94	272	58	11.88	88	27	5.264
Saturn·	2	29	40.190	348	0	50.68	337	45	52.32	10	13	49.89
Uranus	0	46	20.540	27	7	31.73	95	0	58.70	292	6	31.40
Neptune	1	47	1.680	330	14	7.90	273	9	58.47	57	4	3.92

Earth by—	K′			log k	log k′	log C	resid.
	°	′	″				″
Mercury	25	3	36.28	9.9992608	9.9974965	7.8017097	0.002
Venus	330	51	5.12	9.9994999	9.9997387	5.3891826	0.008
Mars	127	4	14.72	9.9997885	9.9999850	8.3052599	0.009
Jupiter	88	27	10.859	9.9998865	9.9999998	8.7995614	0.010
Saturn	10	16	6.88	9.9999411	9.9996473	9.4563012	0.003
Uranus	292	6	34.66	9.9999609	9.9999997	9.9089914	0.005
Neptune	57	4	14.94	9.9997902	9.9999994	8.8147322	0.001

Mars by—	I			П			П′			K		
	°	′	″	°	′	″	°	′	″	°	′	″
Mercury	5	9	10.165	287	24	19.31	29	13	54.31	258	16	20.56
Venus	1	56	2.460	232	18	22.07	28	26	43.81	203	52	27.49
Earth	1	51	2.240	104	53	57.15	231	57	45.14	232	57	4.23
Jupiter	1	26	6.381	149	47	4.35	188	22	45.31	321	24	28.37
Saturn	2	21	52.110	176	17	59.42	293	4	38.76	243	12	17.28
Uranus	1	11	40.460	120	39	30.31	315	36	26.40	165	2	41.49
Neptune	2	22	41.388	152	49	56.12	222	47	52.02	290	3	32.71

Mars by—	K′			log k	log k′	log C	resid.
	°	′	″				″
Mercury	258	4	28.68	9.9995819	9.9986621	7.8017097	0.017
·Venus	203	50	49.03	9.9999439	9.9998087	5.3891826	0.012
Earth	232	55	19.79	9.9998596	9.9999141	6.4491252	0.009
Jupiter	321	24	9.72	9.9999971	9.9998667	8.7995614	0.007
Saturn	243	14	23.99	9.9996870	9.9999432	9.4563012	0.007
Uranus	165	3	26.31	9.9999538	9.9999519	9.9089914	0.003
Neptune	290	0	35.50	9.9998274	9.9997986	8.8147322	0.006

9. THE RADII VECTORES AND THE TRUE ANOMALIES.

The values of $\log r$ and v for the points of division employed in the four different cases are given in the following tables. In each case the equations,

$$\Sigma_1 r = \Sigma_2 r = ja, \quad \text{and} \quad \Sigma_1 v + 180° = \Sigma_2 v$$

were exactly satisfied, and the values of r were also obtained from the equation stated in Article 7 for obtaining the value of r^2.

MERCURY.

E	log r	v °	'	"
0	9.4878584	0	0	0.00
15	9.4916716	18	25	28.96
22.5	9.4963313	27	32	14.93
30	9.5026623	36	32	7.50
45	9.5195925	54	4	7.02
60	9.5407098	70	50	41.41
67.5	9.5522314	78	55	7.36
75	9.5640735	86	46	40.73
90	9.5878217	101	51	53.65
105	9.6103385	116	9	54.15
112.5	9.6207149	123	3	1.59
120	9.6303194	129	46	44.60
135	9.6467730	142	49	52.77
150	9.6589887	155	27	29.02
157.5	9.6633518	161	39	20.97
165	9.6664956	167	48	0.75
180	9.6690267	180	0	0.00
195	9.6664956	192	11	59.25
202.5	9.6633518	198	20	39.03
210	9.6589887	204	32	30.98
225	9.6467730	217	10	7.23
240	9.6303194	230	13	15.40
247.5	9.6207149	236	56	58.41
255	9.6103385	243	50	5.85
270	9.5878217	258	8	6.35
285	9.5640735	273	13	19.27
292.5	9.5522314	281	4	52.64
300	9.5407098	289	9	18.59
315	9.5195925	305	55	52.98
330	9.5026623	323	27	52.50
337.5	9.4963313	332	27	45.08
345	9.4916716	341	34	31.04

VENUS.

E	log r	v °	'	"
0	9.8563557	0	0	0.00
15	9.8564576	15	6	6.54
30	9.8567564	30	11	47.87
45	9.8572313	45	16	40.52
60	9.8578493	60	20	24.50
75	9.8585680	75	22	44.64
90	9.8593378	90	23	31.50
105	9.8601064	105	22	42.20
120	9.8608213	120	20	20.31
135	9.8614342	135	16	35.65
150	9.8619040	150	11	43.65
165	9.8621990	165	6	4.12
180	9.8622996	180	0	0.00
195	9.8621990	194	53	55.88
210	9.8619040	209	48	16.35
225	9.8614342	224	43	24.35
240	9.8608213	239	39	39.69
255	9.8601064	254	37	17.80
270	9.8593378	269	36	28.50
285	9.8585680	284	37	15.36
300	9.8578493	299	39	35.50
315	9.8572313	314	43	19.48
330	9.8567564	329	48	12.13
345	9.8564576	344	53	53.46

THE EARTH.						MARS.				
E	$\log r$	v			E	$\log r$	v		
°		°	′	″	°		°	′	″
0	9.9926546	0	0	0.00	0	0.1403760	0	0	0.00
22.5	9.9932181	22	52	14.25	30	0.1463201	32	47	24.62
30	9.9936460	30	29	2.39	45	0.1532670	48	54	53.41
45	9.9948189	45	41	0.84	60	0.1621567	64	44	46.64
60	9.9963428	60	50	8.59	90	0.1828971	95	21	5.913
67.5	9.9972036	68	23	26.41	120	0.2026920	124	31	47.15
90	0.0000000	90	57	39.46	135	0.2106341	138	39	52.35
112.5	0.0027784	113	23	5.92	150	0.2166313	152	34	23.40
120	0.0036266	120	49	43.50	180	0.2216237	180	0	0.00
135	0.0051200	135	40	31.82	210	0.2166313	207	25	36.60
150	0.0062624	150	28	37.29	225	0.2106341	221	20	7.65
157.5	0.0066776	157	51	53.72	240	0.2026920	235	28	12.85
180	0.0072232	180	0	0.00	270	0.1828971	264	38	54.087
202.5	0.0066776	202	8	6.29	300	0.1621567	295	15	13.36
210	0.0062624	209	31	22.71	315	0.1532670	311	5	6.59
225	0.0051200	224	19	28.18	330	0.1463201	327	12	35.38
240	0.0036266	239	10	16.50					
247.5	0.0027784	246	36	54.08					
270	0.0000000	269	2	20.54					
292.5	9.9972036	291	36	33.59					
300	9.9963428	299	9	51.41					
315	9.9948189	314	18	59.16					
330	9.9936460	329	30	57.61					
337.5	9.9932181	337	7	45.75					

10. THE SEPARATE RESULTS.

The values found for the intermediate auxiliary functions which depend upon E, as well as the final perturbations of the four inner planets in each case are now stated in the following tables. The results of the application of the more important test equations are also shown, but all of the test equations of Article 7 were also applied, and each computation (except the first), was, after its completion, duplicated from the beginning, the forms of the equations being changed in the duplication when this was possible.

MERCURY.

ACTION OF VENUS ON MERCURY.

E	A	$B \cos \epsilon$	$B \sin \epsilon$	$1000000 \times g$	h
0°	0.619543952	+ 0.13308441	− 0.18036925	0.7970904	0.5235861$\overline{4}$
30	0.627434998	+ 0.22218381	− 0.06982371	0.1194506	0.5239083$\overline{6}$
60	0.647116316	+ 0.24372756	+ 0.07193966	0.1268000	0.52384406
90	0.675632886	+ 0.19194286	+ 0.20693555	0.0491867	0.52344851
120	0.706503003	+ 0.08070542	+ 0.29899200	2.1902889	0.52319742
150	0.730295757	− 0.06017874	+ 0.32344233	1.5631633	0.52358280
180	0.738317327	− 0.19295989	+ 0.27373528	1.8358797	0.52446104
210	0.727259050	− 0.28205939	+ 0.16318979	0.6524819	0.5250077$\overline{8}$
240	0.701243272	− 0.30360314	+ 0.02142638	0.0112481	0.5247075$\overline{5}$
270	0.669559472	− 0.25181838	− 0.11356958	0.3160138	0.5239107$\overline{5}$
300	0.641856586	− 0.14058090	− 0.20562585	1.0359483	0.52329644
330	0.624398293	+ 0.00030325	− 0.23007624	1.2969588	0.52323374
Σ_1	4.054580456*	− 0.17962654†	+ 0.28009822‡	5.9972554	3.14309264
Σ_2	.4.054580456	− 0.17962659	+ 0.28009814	5.9972551	3.1430919$\overline{3}$

E	l	G	G'	G''	θ		
					°	′	″
0	0.0959333$\overline{2}$	0.5235825$\overline{8}$	0.09595274	0.000015866	25	20	53.90
30	0.1035021$\overline{5}$	0.5239078$\overline{2}$	0.10350489	0.000002203	26	23	25.33
60	0.12324776	0.5238434$\overline{6}$	0.12325032	0.000001964	29	0	59.15
90	0.15215988	0.52344311	0.15217844	0.000013171	32	37	46.70
120	0.1832810$\overline{9}$	0.52318510	0.18331625	0.000022837	36	17	45.71
150	0.20668846	0.52356735	0.20672760	0.000023681	38	55	52.70
180	0.21383179	0.52444977	0.21385942	0.000016369	39	41	12.31
210	0.2022267$\overline{8}$	0.5250039$\overline{3}$	0.20223677	0.000006145	38	21	51.31
240	0.1765112$\overline{3}$	0.5247074$\overline{9}$	0.17651140	0.000000121	35	27	1.91
270	0.1456242$\overline{3}$	0.52390915	0.14562996	0.000004142	31	49	7.06
300	0.11853565	0.52329155	0.11855723	0.000016698	28	25	30.42
330	0.10114005	0.52322787	0.10117042	0.000024501	26	5	20.70
Σ_1	0.9113408$\overline{3}$	3.14305994	0.91144736	0.000073855	194	13	23.40
Σ_2	0.9113415$\overline{4}$	3.14305922	0.91144808	0.000073843	194	13	23.80

* $6a^2 + 3a^2e^2 + 6[a'^2 - 2kaa'ee' \cos K] = + 4.054580460.$
† $6[a'^2e' - kaa'e \cos K] = - 0.17962650.$
‡ $- 6k'aa' \cos \varphi' \cdot e \sin K' = + 0.28009816.$

ACTION OF VENUS ON MERCURY.

E	$\log K_0$	$\log L_0'$	$\log N_0$	$\log N$	$\log P$	$\log Q$
0°	0.06678154	0.36107029	0.27485672	9.0518226	9.9748963	9.6076810
30	0.07267844	0.36875602	0.28344481	9.0869397	0.0171823	9.6511278
60	0.08883727	0.38974368	0.30686976	9.1792738	0.1306112	9.7669404
90	0.11429390	0.42259487	0.34345542	9.2994384	0.2842725	9.9240134
120	0.14429575	0.46098687	0.38608356	9.4147448	0.4383831	$0.082154\overline{1}$
150	0.16872258	0.49199359	0.42040790	$9.496033\overline{5}$	$0.550042\overline{8}$	0.1974492
180	0.17620114	0.50144271	0.43084933	9.5225000	0.5845077	0.2336318
210	0.16325170	0.48506821	0.41274966	$9.488799\overline{4}$	$0.533532\overline{3}$	$0.181381\overline{4}$
240	0.13698082	0.45165804	0.37573831	9.4055654	$0.417388\overline{9}$	0.0613864
270	0.10823963	0.41480523	0.33478930	9.2928156	0.2691020	9.9083455
300	0.08503270	0.38481172	0.30136861	$9.176137\overline{7}$	0.1234343	9.7587487
330	0.07094409	0.36649704	0.28092115	9.0860236	$0.015098\overline{3}$	$9.648233\overline{6}$
Σ_1	0.69812922	2.54971331	2.07576629	$5.750044\overline{3}$	$1.669221\overline{5}$	$9.510542\overline{4}$
Σ_2	0.69813034	2.54971496	2.07576824	5.7500501	$1.669230\overline{1}$	9.5105508

E	$\log V$	J_1'	$1000 \times J_2$	J_3	$1000 \times F_2$
0°	$9.607665\overline{0}$	0.521404654	− 2.7049984	− 0.024195167	+ 0.6439191
30	9.6511256	0.520191194	− 0.6256166	− 0.032354328	+ 0.2492710
60	9.7669384	0.521003862	+ 1.8268277	− 0.030122813	− 0.2568249
90	9.9240003	0.522559008	+ 2.6401451	− 0.018096342	− 0.7387609
120	$0.082131\overline{6}$	0.523207843	+ 2.0172777	+ 0.000503667	− 1.0674025
150	0.1974260	0.522626872	+ 1.0213969	+ 0.020692271	− 1.1546906
180	$0.233615\overline{9}$	0.521405157	+ 0.3980040	+ 0.037057786	− 0.9772364
210	$0.181375\overline{4}$	0.520383901	+ 0.3750436	+ 0.045213990	− 0.5825883
240	0.0613863	0.520288911	+ 0.7059590	+ 0.042976535	− 0.0764923
270	9.9083414	0.521364846	+ 0.6938449	+ 0.030947096	+ 0.4054437
300	9.7587319	0.522844254	− 0.4295092	+ 0.012350057	+ 0.7340853
330	9.6482088	0.523030858	− 2.2812182	− 0.007832614	+ 0.8213734
Σ_1	$9.510469\overline{0}$	3.130154681*	+ 1.8135608	+ 0.038570065	− 0.9999517
Σ_2	$9.510477\overline{5}$	3.130156679	+ 1.8235957	+ 0.038570073	− 0.9999517

* $\Sigma_1(J_1' - G'') = + 3.130080826.$
$\Sigma_2(J_1' - G'') = + 3.130082836.$

ACTION OF VENUS ON MERCURY.

E	$10000 \times F_2$	R_0	$1000 \times S_0$	W_0	$R^{(n)}$	$1000 \times S^{(n)}$
0°	− 0.24640136	0.05971623	− 0.4883004	− 0.009827036	0.00000000	− 1.5879204
30	− 0.00762098	0.06444673	− 0.0208471	− 0.014490447	+ 0.10127648	− 0.0655216
60	− 0.09149188	0.08146579	+ 0.7212240	− 0.017625381	+ 0.20314033	+ 2.0766362
90	− 0.49802851	0.11164771	+ 0.7947010	− 0.015287000	+ 0.28842180	+ 2.0529668
120	− 0.89441990	0.15077545	− 0.4917115	+ 0.000363093	+ 0.30587345	− 1.1518366
150	− 0.92808606	0.18433784	− 2.4881611	+ 0.032271915	+ 0.20211372	− 5.4561938
180	− 0.56751805	0.19359808	− 3.0725539	+ 0.063241297	0.00000000	− 6.5837414
210	− 0.13320998	0.17488253	− 1.4207354	+ 0.068605698	− 0.19174661	− 3.1154764
240	+ 0.01209827	0.14154301	+ 0.6131517	+ 0.049504580	− 0.28714394	+ 1.4363109
270	− 0.19301611	0.10799082	+ 1.3152406	+ 0.025022994	− 0.27897487	+ 3.3976883
300	− 0.46971306	0.08194590	+ 0.7289596	+ 0.007023584	− 0.20433749	+ 2.0989094
330	− 0.49748432	0.06575439	− 0.1643583	− 0.003535808	− 0.10333144	− 0.5165700
Σ_1	− 2.25744598	0.70904446	− 1.9892305	+ 0.092680137	+ 0.01753235	− 3.7116419
Σ_2	− 2.25744596	0.70906002	− 1.9841603	+ 0.092587352	+ 0.01775908	− 3.7031067

E	$R_0 \sin v$ $+ (\cos v + \cos E)S_0$	$- R_0 \cos v$ $+ \left(\dfrac{r}{a}\sec^2 \varphi +1\right)\sin v \cdot S_0$	$W_0 \cos u$	$W_0 \sin u$	$- 2\dfrac{r}{a} R_0$
0°	− 0.00097660	− 0.05971623	− 0.008630594	− 0.004699311	− 0.09487655
30	+ 0.03833160	− 0.05180530	− 0.006100201	− 0.013143840	− 0.10594280
60	+ 0.07755258	− 0.02541165	+ 0.002882594	− 0.017388064	− 0.14618182
90	+ 0.10909895	+ 0.02454507	+ 0.009914505	− 0.011635936	− 0.22329541
120	+ 0.11643405	+ 0.09565744	− 0.000337472	+ 0.000133975	− 0.33255092
150	+ 0.08098459	+ 0.16537954	− 0.032192272	− 0.002265848	− 0.43432170
180	+ 0.00614511	+ 0.19359808	− 0.055541660	− 0.030242129	− 0.46680548
210	− 0.07011629	+ 0.16039915	− 0.041182679	− 0.054870101	− 0.41204381
240	− 0.10947718	+ 0.08954949	− 0.009624818	− 0.048559933	− 0.31218788
270	− 0.10595403	+ 0.01957233	+ 0.007191939	− 0.023967193	− 0.21598164
300	− 0.07680517	− 0.02822239	+ 0.005196776	− 0.004724857	− 0.14704334
330	− 0.03941900	− 0.05265111	− 0.003501678	+ 0.000490095	− 0.10809245
Σ_1	+ 0.01287279	+ 0.26545474	− 0.066055174	− 0.105480319	− 1.49964599
Σ_2	+ 0.01292582	+ 0.26543968	− 0.065870386	− 0.105392823	− 1.49967781

$$\sin \varphi \cdot \tfrac{1}{2}A_1^{(s)} + \cos \varphi \cdot B_0^{(c)} = - 0.0000000083.$$

DIFFERENTIAL COEFFICIENTS.

			log coeff.
$[de/dt]_{00}$	$= +$	$11321.398\ m'$	$p\ 4.0539001$
$[d\chi/dt]_{00}$	$= +$	$1133127.6\ \ m'$	$p\ 6.0542788$
$[di/dt]_{00}$	$= -$	$60449.278\ m'$	$n\ 4.7813911$
$[d\Omega/dt]_{00}$	$= -$	$792605.00\ m'$	$n\ 5.8990568$
$[d\pi/dt]_{00}$	$= +$	$1127216.0\ \ m'$	$p\ 6.0520072$
$[dL/dt]_{00}$	$= -$	$1326653.0\ \ m'$	$n\ 6.1227573$

FINAL VALUES CORRESPONDING TO THE ABOVE VALUE OF m'.

$$[de/dt]_{00} = +0.027739414$$
$$[d\chi/dt]_{00} = +2.7763615$$
$$[di/dt]_{00} = -0.14811133$$
$$[d\Omega/dt]_{00} = -1.9420214$$
$$[d\pi/dt]_{00} = +2.7618772$$
$$[dL/dt]_{00} = -3.2505323$$

COMPARISON WITH OTHER RESULTS.

	Leverrier.	Newcomb.	Hill.	Method of Gauss.
$[de/dt]_{00}$	$+0.02780$	$+0.02774$	$+0.0277391$	$+0.0277394$
$e[d\pi/dt]_{00}$	$+0.56811$	$+0.57086$	$+0.567852$	$+0.567855$
$[di/dt]_{00}$	-0.14812	-0.14806	-0.1481112	-0.1481113
$\sin i\,[d\Omega/dt]_{00}$	-0.23648	-0.23665	-0.2367447	-0.2367449
$[d\chi/dt]_{00}$			$+2.776347$	$+2.776361$
$[dL/dt]_{00}$	-3.2769		-3.250522	-3.250532

NOTES.

This is the only one of the twenty eight computations that was not duplicated, but the values of θ were computed by two different formulas and all known test equations were applied. As an illustration of his first modification of GAUSS's method HILL published this complete computation from exactly the same elements as here employed, and DR. LOUIS ARNDT states that he has verified the results and found them correct. (*Bulletin de la Societe des Sciences Naturelles de Neuchatel,* Vol. XXIV). INNES states however that the test arising from the constancy of the major axis is not satisfied, the residual being -0.00075 (*M. N.,* Vol. LII, page 87), but this statement is an error, for the residual obtained from HILL's figures is -0.0000000088, a practically exact agreement with that here obtained.

Upon comparing the present computation with that of HILL, the following slight discrepancies may be noticed:

Π', K and K' differ by less than $0''.1$ from HILL's values, a difference doubtless due to the fact that the preliminary computation was here effected with eight place logarithms while HILL employed but seven. The value of l for $330°$ should be 0.10114009 instead of 0.11014009, and G'' for $180°$ should be 0.00001637 instead of 0.00001617. These are misprints merely. The values of the logarithms of K_0, L_0', and N_0 in HILL seem to be slightly in error throughout, a double interpolation to second differences from HILL's values of θ giving with the three functions most in error,

Hill's Values.

For $E = 60°$,	$\log N_0$	0.30686978	0.3068691
For $E = 150°$,	$\log L_0'$	0.49199342	0.4919942
For $E = 180°$,	$\log L_0'$	0.50144261	0.5014421

The effect of these differences upon the final coefficients is, however, almost inappreciable.

It is evident from an inspection of the final sums that a division into twelve parts is necessary in this case, the terms from the sixth to the eleventh orders, inclusive, amounting to 1/600th of the whole for $[di/dt]_{00}$ and to 1/1200th of the whole for $[d\Omega/dt]_{00}$.

ACTION OF THE EARTH ON MERCURY.

E	A	$B \cos \epsilon$	$B \sin \epsilon$	$10000 \times g$	h
0°	1.10386215	$+ 0.29403604$	$- 0.13175730$	0.04882864	1.00008277
30	1.11164085	$+ 0.32704393$	$+ 0.06103685$	0.01047875	1.00111148
60	1.12870093	$+ 0.25768623$	$+ 0.24661358$	0.17106421	1.00173668
90	1.15278960	$+ 0.10454706$	$+ 0.37524793$	0.39606073	1.00128597
120	1.17861164	$- 0.09134000$	$+ 0.41247210$	0.47853578	1.00022588
150	1.19808875	$- 0.27748723$	$+ 0.34831240$	0.34124242	0.99978030
180	1.20368356	$- 0.40401673$	$+ 0.19995995$	0.11246339	1.00066133
210	1.19273750	$- 0.43702459$	$+ 0.00716576$	0.00014443	1.00216106
240	1.16934301	$- 0.36766690$	$- 0.17841099$	0.08952996	1.00285589
270	1.14208711	$- 0.21452774$	$- 0.30704518$	0.26517354	1.00212493
300	1.11943230	$- 0.01864062$	$- 0.34426952$	0.33336685	1.00067710
330	1.10628960	$+ 0.16750659$	$- 0.28010968$	0.22068934	0.99977580
Σ_1	6.90363359*	$- 0.32994198†$	$+ 0.20460782‡$	1.23378883	6.00623965
Σ_2	6.90363341	$- 0.32994198$	$+ 0.20460808$	1.23378921	6.00623954

* $6a^2 + 3a^2e^2 + 6[a'^2 - 2kaa'ee' \cos K] = + 6.90363352$.

† $6[a'^2e' - kaa'e \cos K] = - 0.32994198$.

‡ $- 6k'aa' \cos \varphi' \cdot e \sin K' = + 0.20460788$.

ACTION OF THE EARTH ON MERCURY.

E	l	G	G'	G''	θ		
°					°	′	″
0	0.10349811	1.00007732	0.10355071	0.00004715	18	46	28.61
30	0.11024810	1.00111031	0.11025877	0.00000949	19	22	58.72
60	0.12668299	1.00171716	0.12683714	0.00013464	20	51	17.40
90	0.15122236	1.00123943	0.15152995	0.00026105	22	54	42.04
120	0.17810449	1.00016768	0.17843084	0.00026815	25	0	4.02
150	0.19802718	0.99973772	0.19824193	0.00017219	26	27	9.31
180	0.20274096	1.00064732	0.20281046	0.00005541	26	45	34.58
210	0.19029517	1.00216104	0.19029526	0.00000008	25	50	0.71
240	0.16620585	1.00284522	0.16627021	0.00005369	24	1	53.32
270	0.13968091	1.00209425	0.13990074	0.00018915	21	57	13.77
300	0.11847394	1.00063933	0.11879216	0.00028045	20	10	34.78
330	0.10623253	0.99975109	0.10646458	0.00020734	19	3	48.08
Σ_1	0.89570634	6.00609403	0.89669152	0.00083949	135	35	52.71
Σ_2	0.89570625	6.00609384	0.89669123	0.00083930	135	35	52.63

E	$\log K_0$	$\log L_0'$	$\log N_0$	$\log N$	$\log P$	$\log Q$
0°	0.03586144	0.32053269	0.22947519	8.5993188	8.9197434	8.8287400
30	0.03828768	0.32372814	0.23305754	8.6307049	8.9534609	8.8632763
60	0.04451494	0.33191845	0.24223558	8.7125510	9.0428624	8.9539831
90	0.05408514	0.34447345	0.25629369	8.8165734	9.1597444	9.0722159
120	0.06487989	0.35858872	0.27208270	8.9130565	9.2712667	9.1849500
150	0.07303960	0.36922627	0.28397010	8.9788974	9.3482020	9.2629066
180	0.07483693	0.37156571	0.28658308	8.0002544	9.3712099	9.2865324
210	0.06949181	0.36460451	0.27880654	8.9738846	9.3366140	9.2517535
240	0.05973313	0.35186484	0.26456369	8.9063079	9.2556584	9.1696145
270	0.04950056	0.33846383	0.24956627	8.8114799	9.1479627	9.0600557
300	0.04158092	0.32806159	0.23791429	8.7102233	9.0374863	8.9477383
330	0.03700237	0.32203558	0.23116016	8.6301758	8.9522476	8.8613541
Σ_1	0.32140725	2.06253200	1.53285453	2.8417119	4.8982271	4.3715583
Σ_2	0.32140716	2.06253178	1.53285430	2.8417160	4.8982316	4.3715621

ACTION OF THE EARTH ON MERCURY.

E	$\log V$	J_1'	J_2	J_3	$1000 \times F_2$
0°	8.8287147	0.9963685	− 0.0084115617	− 0.04554861	+ 2.1929308
30	8.8632712	0.9875038	− 0.0046146363	− 0.08975347	− 1.0158793
60	8.9539113	0.9853935	+ 0.0065532595	− 0.10655243	− 4.1045657
90	9.0720772	0.9913722	+ 0.0136249408	− 0.09143424	− 6.2455186
120	9.1848079	0.9979641	+ 0.0119853300	− 0.04844492	− 6.8650672
150	9.2628156	0.9998176	+ 0.0047535770	+ 0.01089164	− 5.7972107
180	9.2865032	0.9963768	− 0.0029434883	+ 0.07066640	− 3.3280756
210	9.2517534	0.9902153	− 0.0070671940	+ 0.11485779	− 0.1192649
240	9.1695861	0.9854773	− 0.0058653024	+ 0.13162975	+ 2.9694212
270	9.0599551	0.9862780	− 0.0010764519	+ 0.11649805	+ 5.1103718
300	8.9475885	0.9932756	+ 0.0016290397	+ 0.07352222	+ 5.7299237
330	8.8612431	0.9996406	− 0.0026304325	+ 0.01421265	+ 4.6620645
Σ_1	4.3711117	5.9548558*	+ 0.0029472768	+ 0.08527241	− 3.4054328
Σ_2	4.3711156	5.9548275	+ 0.0029898031	+ 0.08527242	− 3.4054372

	$1000 \times F_3$	R_0	$1000 \times S_0$	$1000 \times W_0$	$R^{(n)}$	$1000 \times S^{(n)}$
0°	− 0.18791333	0.020434768	− 0.3847186	− 3.085993	0.000000000	− 1.2510793
30	+ 0.04317409	0.021303927	− 0.4280932	− 6.547345	+ 0.033478638	− 1.3454762
60	− 0.07397385	0.025621038	+ 0.1363110	− 9.590571	+ 0.063887706	+ 0.3925293
90	− 0.49750230	0.033590134	+ 0.7062450	− 10.865960	+ 0.086774000	+ 1.8244563
120	− 0.86425140	0.043556201	+ 0.5521780	− 7.575463	+ 0.088361000	+ 1.2934800
150	− 0.83674077	0.051546760	− 0.4218487	+ 1.808294	+ 0.056517506	− 0.9250562
180	− 0.43280690	0.053431000	− 1.3516860	+ 13.566615	0.000000000	− 2.8963373
210	− 0.01035886	0.048659649	− 1.2877184	+ 20.505307	− 0.053351926	− 2.8237889
240	− 0.07819780	0.040457621	− 0.3317444	+ 19.465045	− 0.082075038	− 0.7771130
270	+ 0.18040025	0.032377496	+ 0.5949004	+ 13.349022	− 0.083641596	+ 1.5368184
300	− 0.46718624	0.026190705	+ 0.7690320	+ 6.465459	− 0.065308303	+ 2.2142904
330	− 0.46610559	0.022213151	+ 0.2265587	+ 0.990810	− 0.034907444	+ 0.7120630
Σ_1	− 1.94793392	0.209691333	− 0.6106280	+ 19.245092	+ 0.004865365	− 1.0242299
Σ_2	− 1.94793368	0.209691117	− 0.6099562	+ 19.240128	+ 0.004869178	− 1.0209836

* $\Sigma_1(J_1' - G'') = 5.9540163.$

$\Sigma_2(J_1' - G'') = 5.9539882.$

ACTION OF THE EARTH ON MERCURY.

E	$\sin v \cdot R_0$ $+ (\cos v + \cos E)S_0$	$- \cos v \cdot R_0$ $+ \left(\dfrac{r}{a}\sec^2 \varphi + 1\right)\sin v S_0$	$100 \times W_0 \cos u$	$100 \times W_0 \sin u$	$- 2\dfrac{r}{a} R_0$
0°	− 0.000769437	− 0.020434774	− 0.2710273	− 0.1475729	− 0.03246657
30	+ 0.011967951	− 0.017591057	− 0.2756307	− 0.5938896	− 0.03502118
60	+ 0.024315363	− 0.008157573	− 0.1568518	− 0.9461439	− 0.04597425
90	+ 0.032727249	+ 0.008319102	+ 0.7047203	− 0.8270793	− 0.06718020
120	+ 0.032844271	+ 0.028781472	+ 0.7040911	− 0.2795215	− 0.09606764
150	+ 0.022159505	+ 0.046499174	− 0.1803831	− 0.0126962	− 0.12145034
180	+ 0.002703372	+ 0.053431012	− 1.1914874	− 0.6487582	− 0.12883338
210	− 0.017924683	+ 0.045456375	− 1.2308941	− 1.6399928	− 0.11464781
240	− 0.030714227	+ 0.026434445	− 0.3784440	− 1.9093573	− 0.08923343
270	− 0.031808132	+ 0.005466897	+ 0.3836684	− 1.2785779	− 0.06475510
300	− 0.024103783	− 0.010000896	+ 0.4783818	− 0.4349396	− 0.04699652
330	− 0.012845686	− 0.018098650	+ 0.0981246	− 0.0137335	− 0.03651581
Σ_1	+ 0.004275559	+ 0.070053686	− 0.5016340	− 4.3662934	− 0.43957179
Σ_2	+ 0.004276204	+ 0.070051841	− 0.5003946	− 4.3659693	− 0.43957044

$$\sin \varphi \cdot \tfrac{1}{2}A_1{}^{(s)} + \cos \varphi \cdot B_0{}^{(c)} = - 0.00000000016.$$

DIFFERENTIAL COEFFICIENTS.

<table>
<tr><td></td><td></td><td>''</td><td></td><td>log coeff.</td></tr>
<tr><td>$[de/dt]_{00}$</td><td>=</td><td>+3752.8345</td><td>m'</td><td>p 3.5743594</td></tr>
<tr><td>$[d\chi/dt]_{00}$</td><td>=</td><td>+299037.72</td><td>m'</td><td>p 5.4757260</td></tr>
<tr><td>$[di/dt]_{00}$</td><td>=</td><td>−4591.3713</td><td>m'</td><td>n 3.6619424</td></tr>
<tr><td>$[d\Omega/dt]_{00}$</td><td>=</td><td>−328217.95</td><td>m'</td><td>n 5.5161623</td></tr>
<tr><td>$[d\pi/dt]_{00}$</td><td>=</td><td>+296589.74</td><td>m'</td><td>p 5.4721561</td></tr>
<tr><td>$[dL/dt]_{00}$</td><td>=</td><td>−390282.17</td><td>m'</td><td>n 5.5913787</td></tr>
</table>

FINAL VALUES CORRESPONDING TO THE ABOVE VALUE OF m'.

$$[de/dt]_{00} = +0.011476557\,{''}$$
$$[d\chi/dt]_{00} = +0.91448833$$
$$[di/dt]_{00} = -0.014040890$$
$$[d\Omega/dt]_{00} = -1.0037245$$
$$[d\pi/dt]_{00} = +0.90700208$$
$$[dL/dt]_{00} = -1.1935233$$

COMPARISON WITH OTHER RESULTS.

	Leverrier.	Newcomb.	Method of Gauss.
$[de/dt]_{00}$	$+0.01153$	$+0.01147$	$+0.0114766$
$e[d\pi/dt]_{00}$	$+0.18658$	$+0.18799$	$+0.186484$
$[di/dt]_{00}$	-0.01414	-0.01404	-0.0140409
$\sin i \, [d\Omega/dt]_{00}$	-0.12219	-0.12233	-0.122360
$[dL/dt]_{00}$	-1.1942		-1.19352

NOTES.

As a' and e' are both small in this case, the sums, up to and including R_0, are in very exact agreement. But as I and e are unusually large, the final sums differ considerably, the greatest discrepancy being in $W_0 \cos u$, which shows that a neglect of the terms from the 6th to the 11th orders would produce an error in $[di/dt]_{00}$ of slightly more than 1/1000th of the whole value of this coefficient.

ACTION OF MARS ON MERCURY.

E	A	$B \cos \epsilon$	$B \sin \epsilon$	g	h
0°	2.3984504	+ 0.12138918	+ 0.45632916	0.0042055	2.3024514
15	2.3737047	− 0.02025060	+ 0.40611421	0.0033309	2.3025214
30	2.3556032	− 0.14407564	+ 0.32017434	0.0020703	2.3025321
45	2.3456111	− 0.24164776	+ 0.20436670	0.0008435	2.3024889
60	2.3447258	− 0.30631712	+ 0.06658267	0.0000896	2.3024184
75	2.3533241	− 0.33367707	− 0.08378742	0.0001418	2.3023625
90	2.3710514	− 0.32186246	− 0.23649650	0.0011296	2.3023658
105	2.3967846	− 0.27167903	− 0.38113729	0.0029337	2.3024640
120	2.4286852	− 0.18654648	− 0.50785318	0.0052088	2.3026710
135	2.4643476	− 0.07226635	− 0.60800831	0.0074658	2.3029751
150	2.5010251	+ 0.06337314	− 0.67477750	0.0091956	2.3033343
165	2.5359018	+ 0.21112855	− 0.70361065	0.0099983	2.3036852
180	2.5663693	+ 0.36093056	− 0.69254258	0.0096862	2.3039570
195	2.5902662	+ 0.50257031	− 0.64232765	0.0083325	2.3040870
210	2.6060491	+ 0.62639549	− 0.55638795	0.0062519	2.3040409
225	2.6128739	+ 0.72396747	− 0.44058010	0.0039202	2.3038234
240	2.6105920	+ 0.78863682	− 0.30279611	0.0018517	2.3034773
255	2.5996753	+ 0.81599664	− 0.15242602	0.0004692	2.3030742
270	2.5810993	+ 0.80418226	+ 0.00028298	0.0000000	2.3026229
285	2.5562149	+ 0.76399902	+ 0.14492387	0.0004242	2.3023962
300	2.5266327	+ 0.66886612	+ 0.27163969	0.0014902	2.3022243
315	2.4941374	+ 0.55458616	+ 0.37179487	0.0027917	2.3021798
330	2.4606272	+ 0.41894655	+ 0.43856414	0.0038844	2.3022350
345	2.4280691	+ 0.27119119	+ 0.46739720	0.0044120	2.3023425
Σ_1	29.7509107*	+ 2.89391842†	− 1.41728084‡	0.0450638	27.6343997
Σ_2	29.7509107	+ 2.89391853	− 1.41728059	0.0450638	27.6344002

* $12a^2 + 6a^2e^2 + 12[a'^2 - 2kaa'ee' \cos K] = 29.7509106.$

† $12[a'^2e' - kaa'e \cos K] = + 2.89391844.$

‡ $- 12k'aa' \cos \varphi' \cdot e \sin K' = - 1.41728062.$

ACTION OF MARS ON MERCURY.

E	l	G	G'	G''	θ		
°					°	′	″
0	0.0758032	2.3016305	0.0957141	0.0190900	12	51	3.65
15	0.0509875	2.3018785	0.0717874	0.0201570	11	28	40.15
30	0.0328754	2.3021359	0.0509293	0.0176577	9	54	3.88
45	0.0229265	2.3023282	0.0338958	0.0108086	7	59	27.93
60	0.0221116	2.3024014	0.0237650	0.0016363	6	1	37.49
75	0.0307659	2.3023354	0.0326775	0.0018845	7	2	5.28
90	0.0484898	2.3021481	0.0572743	0.0085668	9	43	5.24
105	0.0741248	2.3018919	0.0890147	0.0143178	12	11	37.36
120	0.1058185	2.3016404	0.1249596	0.0181104	14	22	46.33
135	0.1411768	2.3014735	0.1626257	0.0199473	16	17	11.47
150	0.1774951	2.3014531	0.1994129	0.0200367	17	54	21.70
165	0.2120208	2.3016063	0.2327627	0.0186630	19	13	8.83
180	0.2422166	2.3019140	0.2604177	0.0161582	20	12	26.11
195	0.2659835	2.3023097	0.2806562	0.0128954	20	51	34.72
210	0.2818125	2.3026974	0.2924401	0.0092842	21	10	38.0^8
225	0.2888548	2.3029783	0.2954612	0.0057613	21	10	27.64
240	0.2869190	2.3030785	0.2900893	0.0027715	20	52	41.30
255	0.2764054	2.3029737	0.2772408	0.0007349	20	19	35.05
270	0.2582114	2.3026922	0.2582114	0.0000000	19	33	51.59
285	0.2336230	2.3023071	0.2344977	0.0007857	18	38	24.76
300	0.2042127	2.3019157	0.2076391	0.0031178	17	36	2.08
315	0.1717618	2.3016103	0.1791035	0.0067722	16	29	5.95
330	0.1381965	2.3014548	0.1502129	0.0112362	15	19	13.32
345	0.1055309	2.3014695	0.1221038	0.0156999	14	6	53.42
Σ_1	1.8741623	27.6251620	2.0110657	0.1276658	185	31	50.77
Σ_2	1.8741617	27.6251624	2.0118270	0.1284276	185	48	12.56

ACTION OF MARS ON MERCURY.

E	$\log K_0$	$\log L_0'$	$\log N_0$	$\log N$	$\log P$	$\log Q$
0°	0.01657774	0.29504379	0.20087027	8.0316820	7.5954802	7.8669295
15	0.01319292	0.29055307	0.19582511	8.0355546	7.5943701	7.8655109
30	0.00979383	0.28603832	0.19075139	8.0547661	7.6099058	7.8800682
45	0.00636436	0.28147805	0.18562486	8.0870691	7.6401445	7.9084927
60	0.00361341	0.27781623	0.18150720	8.1291203	7.6819573	7.9481379
75	0.00492729	0.27956556	0.18347444	8.1771100	7.7316276	7.9980604
90	0.00943284	0.28555854	0.19021211	8.2272783	7.7853440	8.0537440
105	0.01490785	0.29282894	0.19838217	8.2762397	7.8395128	8.1098440
120	0.02081742	0.30066160	0.20717937	8.3211159	7.8908949	8.1628540
135	0.02681707	0.30859802	0.21608807	8.3595540	7.9366442	8.2098882
150	0.03254501	0.31616047	0.22457224	8.3896939	7.9743208	8.2484993
165	0.03762579	0.32285663	0.23208058	8.4101312	8.0019112	8.2766735
180	0.04171228	0.32823434	0.23810787	8.4198969	8.0178772	8.2928778
195	0.04453613	0.33194630	0.24226678	8.4184648	8.0212321	8.2961421
210	0.04594748	0.33380024	0.24434354	8.4057702	8.0116018	8.2861294
225	0.04593448	0.33378318	0.24432441	8.3822399	7.9892733	8.2631894
240	0.04461766	0.33205341	0.24238678	8.3488317	7.9552231	8.2283875
255	0.04222050	0.32890264	0.23885673	8.3070780	7.9111256	8.1835072
270	0.03902753	0.32470208	0.23414923	8.2591387	7.8593690	8.1310520
285	0.03533705	0.31984157	0.22870029	8.2078386	7.8030574	8.0742275
300	0.03141955	0.31467566	0.22290683	8.1566450	7.7459663	8.0168746
315	0.02748746	0.30948385	0.21708212	8.1095325	7.6924008	7.9633068
330	0.02367787	0.30444739	0.21142959	8.0706478	7.6468600	7.9179598
345	0.02004728	0.29964170	0.20603415	8.0437755	7.6135016	7.8848519
Σ_1	0.31918262	3.69919207	2.58841642	8.8145868	3.7748004	7.0335140
Σ_2	0.31939818	3.69947951	2.58873971	8.8145879	3.7748012	7.0338946

ACTION OF MARS ON MERCURY.

E	$\log V$	$J_1{}'$	J_2	J_3	F_2
0°	7.8624692	2.3161036	+ 0.034714816	− 0.08283714	− 0.09798270
15	7.8607965	2.3114562	+ 0.028725995	− 0.12496689	− 0.08720058
30	7.8759318	2.3036617	+ 0.023022893	− 0.15748255	− 0.06874762
45	7.9059548	2.2939379	+ 0.017028648	− 0.17817560	− 0.04388148
60	7.9477527	2.2850812	+ 0.009527241	− 0.18564570	− 0.01429659
75	7.9976170	2.2883310	− 0.000387926	− 0.17939386	+ 0.01799079
90	8.0517340	2.2994643	− 0.012768514	− 0.15985316	+ 0.05078038
105	8.1064943	2.3097716	− 0.026802369	− 0.12835823	+ 0.08183755
120	8.1586286	2.3171693	− 0.041120811	− 0.08705254	+ 0.10904591
135	8.2052458	2.3210340	− 0.054165965	− 0.03874380	+ 0.13055114
150	8.2438461	2.3213532	− 0.064480743	+ 0.01328579	+ 0.14488783
165	8.2723467	2.3185110	− 0.070890916	+ 0.06550049	+ 0.15107882
180	8.2891363	2.3131717	− 0.072601961	+ 0.11434922	+ 0.14870229
195	8.2931581	2.3062073	− 0.069243148	+ 0.15650567	+ 0.13792019
210	8.2839812	2.2986444	− 0.060886015	+ 0.18909426	+ 0.11946725
225	8.2618557	2.2916136	− 0.048054879	+ 0.20988694	+ 0.09460109
240	8.2277453	2.2862701	− 0.031734816	+ 0.21745674	+ 0.06501619
255	8.1833367	2.2836646	− 0.013368882	+ 0.21127772	+ 0.03272882
270	8.1310520	2.2845306	+ 0.005188260	+ 0.19176389	− 0.00006076
285	8.0740447	2.2890080	+ 0.021821056	+ 0.16024221	− 0.03111796
300	8.0161484	2.2963713	+ 0.034481515	+ 0.11886354	− 0.05832630
315	7.9617277	2.3049376	+ 0.041716329	+ 0.07045517	− 0.07983156
330	7.9153379	2.3123762	+ 0.043244020	+ 0.01832594	− 0.09416821
345	7.8811861	2.3164879	+ 0.040207969	− 0.03396170	− 0.10035921
Σ_1	7.0037635	27.6341976*	− 0.133414115	+ 0.19026829	+ 0.30431767
Σ_2	7.0037641	27.6349607	− 0.133414088	+ 0.19026812	+ 0.30431761

* $\Sigma_1(J_1{}' - G'') = 27.5065318.$

$\Sigma_2(J_1{}' - G'') = 27.5065331.$

ACTION OF MARS ON MERCURY.

E	F_3	R_0	$1000 \times S_0$	$100 \times W_0$	$R^{(n)}$
0°	+ 0.006125234	0.005413043	− 0.1331172	− 0.05793911	0.000000000
15	+ 0.004244733	0.005395946	− 0.1341971	− 0.08902862	+ 0.004501863
30	+ 0.002091460	0.005581735	− 0.1069853	− 0.11749705	+ 0.008771561
45	+ 0.000394535	0.005978460	− 0.0544826	− 0.14331159	+ 0.012778556
60	− 0.000202379	0.006587462	+ 0.0157366	− 0.16470025	+ 0.016426268
75	+ 0.000673582	0.007397547	+ 0.0931210	− 0.17804909	+ 0.019496561
90	+ 0.003010065	0.008379102	+ 0.1659322	− 0.17823953	+ 0.021645905
105	+ 0.006397644	0.009479745	+ 0.2230373	− 0.15960687	+ 0.022459598
120	+ 0.010124788	0.010622853	+ 0.2557091	− 0.11755697	+ 0.021550254
135	+ 0.013355105	0.011710967	+ 0.2593958	− 0.05060872	+ 0.018676901
150	+ 0.015340435	0.012635664	+ 0.2351775	+ 0.03775311	+ 0.013854131
165	+ 0.015613320	0.013293529	+ 0.1902485	+ 0.13831063	+ 0.007415508
180	+ 0.014107795	0.013605711	+ 0.1366988	+ 0.23722085	0.000000000
195	+ 0.011176655	0.013535474	+ 0.0883172	+ 0.31912559	− 0.007550472
210	+ 0.007500348	0.013096767	+ 0.0561741	+ 0.37133333	− 0.014359702
225	+ 0.003911780	0.012350565	+ 0.0447317	+ 0.38738325	− 0.019696946
240	+ 0.001183526	0.011390146	+ 0.0503246	+ 0.36844987	− 0.023106835
255	− 0.000166297	0.010320547	+ 0.0628133	+ 0.32211263	− 0.024451640
270	+ 0.000001635	0.009239635	+ 0.0697179	+ 0.25931088	− 0.023868989
285	+ 0.001425688	0.008225494	+ 0.0610457	+ 0.19093564	− 0.021678655
300	+ 0.003528129	0.007331827	+ 0.0329172	+ 0.12533212	− 0.018282391
315	+ 0.005583303	0.006590510	− 0.0111915	+ 0.06726176	− 0.014086776
330	+ 0.006923128	0.006018069	− 0.0617561	+ 0.01815027	− 0.009457250
345	+ 0.007124120	0.005623237	− 0.1063107	− 0.02290733	− 0.004691492
Σ_1	+ 0.069734164	0.109902014	+ 0.7165294	+ 0.78161752	− 0.006827048
Σ_2	+ 0.069734168	0.109902021	+ 0.7165286	+ 0.78161728	− 0.006826894

$$\sin \varphi \cdot \tfrac{1}{2}A_1^{(s)} + \cos \varphi \cdot B_0^{(c)} = + 0.000000000104.$$

ACTION OF MARS ON MERCURY.

E	$1000 \times S^{(n}$	$1000 \times [R_0 \sin v + (\cos v + \cos E)S_0]$	$1000 \times \left[-R_0 \cos v + \left(\frac{r}{a}\sec^2\varphi +1\right)\sin vS_0 \right]$	$1000 \times W_0 \cos u$	$1000 \times W_0 \sin u$	$-2\frac{r}{a}R_0$
0°	−0.43288822	−0.2662344	− 5.4130432	−0.5088501	− 0.2770661	−0.008600192
15	−0.43258505	+1.4484906	− 5.1972533	−0.6072560	− 0.6510373	−0.008648634
30	−0.33624961	+3.1443027	− 4.6032128	−0.4946402	− 1.0657799	−0.009175713
45	−0.16468928	+4.7703837	− 3.5917373	−0.1836694	− 1.4212974	−0.010218569
60	+0.04531053	+6.2357682	− 2.1327368	+0.2693639	− 1.6248266	−0.011820509
75	+0.25408211	+7.4151862	− 0.2308954	+0.7622000	− 1.6090993	−0.014007787
90	+0.42865608	+8.1659686	+ 2.0547246	+1.1559864	− 1.3566975	−0.016758205
105	+0.54706506	+8.3522568	+ 4.6004959	+1.3031443	− 0.9215481	−0.019968410
120	+0.59900000	+7.8723845	+ 7.2196098	+1.0926174	− 0.4337649	−0.023429816
135	+0.58504581	+6.6852166	+ 9.6761487	+0.5003962	− 0.0756834	−0.026827124
150	+0.51571179	+4.8307371	+11.7119607	−0.3765995	− 0.0265069	−0.029771109
165	+0.41003953	+2.4394871	+13.0838349	−1.3270492	− 0.3897732	−0.031867221
180	+0.29291257	−0.2733976	+13.6057094	−2.0833919	− 1.1343956	−0.032806220
195	+0.19034846	−3.0319641	+13.1877783	−2.4169344	− 2.0838761	−0.032447216
210	+0.12318210	−5.5396099	+11.8615372	−2.2290420	− 2.9698842	−0.030857525
225	+0.10088867	−7.5290371	+ 9.7823314	−1.5918740	− 3.5316447	−0.028292299
240	+0.11788553	−8.8108894	+ 7.2045350	−0.7163503	− 3.6141908	−0.025122156
255	+0.15406833	−9.3069299	+ 4.4325573	+0.1350559	− 3.2182941	−0.021739497
270	+0.18010328	−9.0565656	+ 1.7602448	+0.7452938	− 2.4836977	−0.018479268
285	+0.16656410	−8.1932578	− 0.5835177	+1.0058653	− 1.6229225	−0.015575563
300	+0.09477911	−6.8986293	− 2.4659985	+0.9273370	− 0.8431253	−0.013156194
315	−0.03382955	−5.3509513	− 3.8502686	+0.6070918	− 0.2895753	−0.011264707
330	−0.19409650	−3.6857779	− 4.7671354	+0.1797507	− 0.0251579	−0.009892996
345	−0.34269706	−1.9808199	− 5.2732726	−0.2254930	− 0.0403424	−0.009012936
Σ_1	+1.43430766	−4.2819430	+36.0361948	−2.0385248	−15.8550934	−0.229869903
Σ_2	+1.43430113	−4.2819391	+36.0362016	−2.0385225	−15.8550938	−0.229869963

DIFFERENTIAL COEFFICIENTS.

log coeff.

$[de/dt]_{00} = - 1879.077 \quad m' \quad n\ 3.2739445$

$[d\chi/dt]_{00} = +76914.75 \quad m' \quad p\ 4.8860096$

$[di/dt]_{00} = - 934.0667 \quad m' \quad n\ 2.9703779$

$[d\Omega/dt]_{00} = -59594.26 \quad m' \quad n\ 4.7752044$

$[d\pi/dt]_{00} = +76470.27 \quad m' \quad p\ 4.8834926$

$[dL/dt]_{00} = -101879.0 \quad m' \quad n\ 5.0080846$

Final Values Corresponding to the Above Values of m'.

$$[de/dt]_{00} = -0.00060742746''$$
$$[d\chi/dt]_{00} = +0.024863343$$
$$[di/dt]_{00} = -0.00030194497$$
$$[d\Omega/dt]_{00} = -0.019264347$$
$$[d\pi/dt]_{00} = +0.024719659$$
$$[dL/dt]_{00} = -0.032933242$$

Comparison with Other Results.

	Leverrier.	Newcomb.	Method of Gauss.
	$''$	$''$	$''$
$[de/dt]_{00}$	-0.00060	-0.00061	-0.000607
$e[d\pi/dt]_{00}$	$+0.00508$	$+0.00511$	$+0.005082$
$[di/dt]_{00}$	-0.00030	-0.00030	-0.000302
$\sin i \, [d\Omega/dt]_{00}$	-0.00234	-0.00235	-0.002348
$[dL/dt]_{00}$	-0.0331		-0.032933

Notes.

On account of the very large values of the eccentricities of both orbits and their high mutual inclination, the approximate test is here wholly inapplicable if but twelve points of division are employed. Thus the two sums differ by $1° 38' 46''.90$ for θ and by $40' 42''.47$ for ϵ, while the sums of the functions immediately dependent upon these quantities differ by proportionate amounts. When the number of points of division is increased to twenty-four, the final sums are in almost exact agreement, showing that the combined effect of all terms from the 11th to the 23rd orders is wholly inappreciable. The greatest difference which arises in the variations from the employment of twenty-four points of division, instead of twelve, occurs in the case of $[di/dt]_{00}$ and here produces a decrease of but three units in the seventh decimal of the logarithm of the coefficient.

ACTION OF JUPITER ON MERCURY.

E	A	$B \cos \epsilon$	$B \sin \epsilon$	p	q^2
0°	27.23340536	+2.0282403	+1.4219711	9.05679111	80.552426
30	27.14356714	+1.0282450	+1.6206004	9.02684503	80.826371
60	27.06879996	+0.0526625	+1.2863778	9.00192264	81.050843
90	27.03145602	−0.6371012	+0.5088565	8.98947466	81.156986
120	27.04270097	−0.8562240	−0.5036265	8.99322298	81.117340
150	27.09836241	−0.5459922	−1.4797775	9.01177679	80.952183
180	27.18120744	+0.2104677	−2.1580381	9.03939180	80.714558
210	27.26787830	+1.2104630	−2.3566674	9.06828209	80.466952
240	27.33631108	+2.1860457	−2.0224444	9.09109301	80.265977
270	27.37048776	+2.8758091	−1.2449232	9.10248524	80.156946
300	27.36240999	+3.0949315	−0.2324396	9.09979265	80.170254
330	27.31308297	+2.7846997	+0.7437107	9.08335031	80.311910
Σ_1	163.22483480*	+6.7161237†	−2.2081997‡	54.28221419	483.871398
Σ_2	163.22483460	+6.7161234	−2.2082005	54.28221412	483.871348

E	$-\theta'$	g	h	l	G
0°	88° 49′ 5″	0.12745094	27.006742	+0.163630	27.006566
30	89 2 45	0.16554389	27.007491	+0.073044	27.007263
60	89 19 0	0.10430339	27.007569	−0.001801	27.007426
90	89 39 30	0.01632120	27.006899	−0.038475	27.006877
120	89 41 40	0.01598743	27.006246	−0.026577	27.006224
150	89 11 43	0.13801114	27.006455	+0.028875	27.006265
180	88 41 46	0.29354831	27.007536	+0.110638	27.007132
210	·88 20 11	0.35007250	27.008599	+0.196247	27.008116
240	87 48 4	0.25781881	27.008698	+0.264581	27.008341
270	88 19 32	0.09768920	27.007728	+0.299728	27.007593
300	88 22 22	0.00340551	27.006562	+0.292816	27.006557
330	88 35 32	0.03486337	27.006176	+0.243874	27.006128
Σ_1	532° 41′ 57″	0.80250439	162.043353	+0.803287	162.042246
Σ_2	533 9 13	0.80250130	162.043348	+0.803293	162.042242

* $6a^2 + 3a^2 e^2 + 6[a'^2 - 2kaa'ee' \cos K] = +163.22483477.$

† $6[a'^2 e' - kaa'e \cos K] = +6.7161238.$

‡ $-6k'aa' \cos \varphi' \cdot e \sin K' = -2.2082004.$

ACTION OF JUPITER ON MERCURY.

E	G'	G''	θ			$\log K_0$	$\log L_0'$
			°	′	″		
0°	0.188801	0.024995	5	6	8.07	0.0025877	0.2764500
30	0.123075	0.049804	4	35	5.15	0.0020887	0.2757852
60	0.061321	0.062980	3	53	7.75	0.0014995	0.2750001
90	0.011983	0.050435	2	45	10.71	0.0007524	0.2740043
120	0.014441	0.040995	2	35	41.32	0.0006684	0.2738923
150	0.087484	0.058419	4	12	38.17	0.0017613	0.2753489
180	0.173639	0.062597	5	21	37.00	0.0028566	0.2768082
210	0.248823	0.052092	6	3	11.63	0.0036449	0.2778582
240	0.297071	0.032133	6	20	5.30	0.0039929	0.2783216
270	0.311476	0.011613	6	16	40.56	0.0039213	0.2782263
300	0.293250	0.000430	5	59	8.38	0.0035638	0.2777502
330	0.249105	0.005182	5	34	4.63	0.0030827	0.2771094
			°	′	″		
Σ_1	1.028523	0.224130	29	15	47.82	0.0151689	1.6582224
Σ_2	1.031946	0.227545	29	26	50.85	0.0152513	1.6583323

E	$\log N_0$	$\log N$	$\log P$	$\log Q$	$\log V$
0°	0.1799707	6.4183196	3.8310274	5.1664192	5.1659174
30	0.1792229	6.4468140	3.8580378	5.1937562	5.1927566
60	0.1783398	6.5219986	3.9320091	5.2678436	5.2665795
90	0.1772196	6.6157906	4.0252257	5.3607256	5.3597128
120	0.1770937	6.7009450	4.1105921	5.4459161	5.4450927
150	0.1787322	6.7589559	4.1694988	5.5052851	5.5041125
180	0.1803735	6.7800058	4.1918462	5.5278954	5.5266396
210	0.1815544	6.7609474	4.1741434	5.5101707	5.5091259
240	0.1820755	6.7044320	4.1187250	5.4544932	5.4538485
270	0.1819683	6.6198775	4.0347585	5.3701732	5.3699402
300	0.1814329	6.5255910	3.9403890	5.2755478	5.2755392
330	0.1807123	6.4489104	3.8629284	5.1980770	5.1979730
Σ_1	1.0792861	9.6512920	4.1245888	2.1381153	2.1336169
Σ_2	1.0794097	9.6512958	4.1245926	2.1381878	2.1336210

ACTION OF JUPITER ON MERCURY.

E	J_1'	J_2	J_3	F_2	F_3
0°	26.907806	−0.08848287	−1.4438584	−1.8440911	+0.12702123
30	26.754757	−0.00410757	−2.3330959	−2.1016844	+0.05043578
60	26.779325	+0.16212886	−2.4984534	−1.6682462	−0.05353664
90	26.922525	+0.18465069	−1.8956909	−0.6599133	−0.05566955
120	27.032841	+0.04256694	−0.6863554	+0.6531309	+0.07759141
150	27.046752	−0.14516517	+0.8055488	+1.9190575	+0.24215469
180	26.945408	−0.26169947	+2.1803373	+2.7986639	+0.29255865
210	26.802715	−0.24648650	+3.0696706	+3.0562500	+0.18231424
240	26.714620	−0.10777763	+3.2352193	+2.6228182	+0.00934620
270	26.757114	+0.06770763	+2.6325527	+1.6144859	−0.07863175
300	26.911783	+0.13569580	+1.4231210	+0.3014406	−0.02506303
330	27.011072	+0.02690450	−0.0689744	−0.9644853	+0.08731406
Σ_1	161.291783*	−0.11756837	+2.2100104	+2.8637163	+0.42791782
Σ_2	161.294935	−0.11649642	+2.2100109	+2.8637104	+0.42791747

E	$1000 \times R_0$	$1000,000 \times S_0$	$100,000 \times W_0$	$1000 \times R^{(n)}$	$100,000 \times S^{(n)}$
0°	0.12949124	−0.25462237	− 2.1070330	0.00000000	−0.82801500
30	0.13531974	−0.15796961	− 3.6329106	+0.21265173	−0.49649079
60	0.16094783	+0.15688051	− 4.6204172	+0.40133398	+0.45170948
90	0.20322837	+0.35279259	− 4.3458069	+0.52500398	+0.91137646
120	0.25064630	+0.20287634	− 1.9026678	+0.50847837	+0.47523891
150	0.28657259	−0.17989808	+ 2.6073819	+0.31420703	−0.39449173
180	0.29756923	−0.44461029	+ 7.3765067	0.00000000	−0.95269261
210	0.28082350	−0.33962404	+ 9.9405132	−0.30790355	−0.74474862
240	0.24483424	+0.03827880	+ 9.2004732	−0.49668759	+0.08966829
270	0.20310516	+0.33360150	+ 6.1619379	−0.52468566	+0.86179960
300	0.16660105	+0.28219872	+ 2.6818065	−0.41543058	+0.81254093
330	0.14104055	−0.02790130	− 0.1024400	−0.22164184	−0.08769243
Σ_1	1.25008989	−0.01899829	+10.6286684	−0.00230582	+0.04845000
Σ_2	1.25008991	−0.01899894	+10.6286755	−0.00236831	+0.04975249

$$\sin \varphi \cdot \tfrac{1}{2} A_1^{(s)} + \cos \varphi \cdot B_0^{(c)} = + 0.00000000000073.$$

* $\Sigma_1(J_1' - G'') = 161.067653.$
$\Sigma_2(J_1' - G'') = 161.067390.$

Action of Jupiter on Mercury.

E	$1000 \times [R_0 \sin v + (\cos v + \cos E)S_0]$	$1000 \times \left[-R_0 \cos v + \left(\frac{r}{a}\sec^2\varphi +1\right)\sin vS_0 \right]$	$1000 \times W_0 \cos u$	$1000 \times W_0 \sin u$	$1000 \times \left(-2\frac{r}{a}R_0\right)$
$0°$	-0.00509245	-0.12949124	-0.018505013	-0.010075877	-0.20573445
30	$+0.07792118$	-0.11047546	-0.015293859	-0.032953015	-0.22244969
60	$+0.15333585$	-0.04994118	$+0.007556597$	-0.045582053	-0.28880400
90	$+0.19816106$	$+0.04884221$	$+0.028185071$	-0.033078773	-0.40645673
120	$+0.19031365$	$+0.16372520$	$+0.017684089$	-0.007020516	-0.55282671
150	$+0.12222497$	$+0.25901646$	-0.026009473	-0.001830673	-0.67519875
180	$+0.00889221$	$+0.29756923$	-0.064784149	-0.035274618	-0.71750180
210	-0.11061196	$+0.25859897$	-0.059670973	-0.079503145	-0.66165318
240	-0.18859570	$+0.15601907$	-0.017887815	-0.090249100	-0.54000750
270	-0.19945173	$+0.03508579$	$+0.017710220$	-0.059019459	-0.40621028
300	-0.15503995	-0.05982940	$+0.019842790$	-0.018040859	-0.29894808
330	-0.08443003	-0.11301587	-0.001014512	$+0.000141991$	-0.23185401
Σ_1	$+0.00381361$	$+0.37805168$	-0.056093501	-0.206243023	-2.60382254
Σ_2	$+0.00381349$	$+0.37805210$	-0.056093526	-0.206243074	-2.60382263

Differential Coefficients.

$$\begin{aligned}
&&& \text{log coeff.} \\
[de/dt]_{00} &= +\quad 3.3470577\ ''\ m' & & p\ 0.5246632 \\
[d\chi/dt]_{00} &= +1613.8089 & m' & p\ 3.2078521 \\
[di/dt]_{00} &= -\quad 51.404941 & m' & n\ 1.7110049 \\
[d\Omega/dt]_{00} &= -1550.4039 & m' & n\ 3.1904449 \\
[d\pi/dt]_{00} &= +1602.2454 & m' & p\ 3.2047290 \\
[dL/dt]_{00} &= -2312.2863 & m' & n\ 3.3640416
\end{aligned}$$

Final Values Corresponding to the Above Value of m'.

$$\begin{aligned}
[de/dt]_{00} &= +0.00319413\ '' \\
[d\chi/dt]_{00} &= +1.540072 \\
[di/dt]_{00} &= -0.049056191 \\
[d\Omega/dt]_{00} &= -1.4795642 \\
[d\pi/dt]_{00} &= +1.5290366 \\
[dL/dt]_{00} &= -2.2066350
\end{aligned}$$

COMPARISON WITH OTHER RESULTS.

	Leverrier.	Newcomb.	Method of Gauss.
	$''$	$''$	$''$
$[de/dt]_{00}$	$+0.00320$	$+0.00320$	$+0.003194$
$e[d\pi/dt]_{00}$	$+0.31437$	$+0.31664$	$+0.314377$
$[di/dt]_{00}$	-0.04907	-0.04905	-0.049056
$\sin i\,[d\Omega/dt]_{00}$	-0.18042	-0.18037	-0.180368
$[dL/dt]_{00}$	-2.2078		-2.20663

NOTES.

The above results were published in 1896 in *A. J.*, No. 386. In 1911, upon applying to the various computations all of the test equations devised or learned of by that time, a slight error was detected in the value of F_3 for 240°. This rendered the values of W_0, $W_0 \cos u$, $W_0 \sin u$, $[di/dt]_{00}$ and $[d\Omega/dt]_{00}$ incorrect.

In this computation the device was for the first time applied of finding the root G by approximations and then depressing the cubic equation and solving the resulting quadratic equation directly. When a' and hence g is large, some such device becomes necessary as the solution by HILL's formulas involves a great amount of labor. Thus, while but three approximations were necessary with the Earth on Mercury, no less than eleven were required in some cases with Mars on Mercury, and in the latter, as well as in the present case, if the formulas of HILL's second method are employed the angle θ' will be found so nearly equal to 90° as to render the values of the roots obtained from it but little better than first approximations. Accordingly all the remaining computations have been effected by the method here outlined, a method which, since the approximation to the value of G is always very rapid, leads so quickly to the values of the roots that special devices for avoiding the solution of the cubic seem unnecessary.

The final sums are here practically in exact agreement, showing that the effect of all terms from the sixth to the eleventh orders inclusive is inappreciable.

.THE SECULAR VARIATIONS OF THE ELEMENTS

ACTION OF SATURN ON MERCURY.

E	A	$B \cos \epsilon$	$B \sin \epsilon$	g	h
0°	91.40055452	7.9236831	−0.7566872	0.1637322	90.704247
22.5	91.41316673	8.0026121	+0.6479546	0.1200578	90.704730
45	91.37113281	7.5285556	+1.9837667	1.1253368	90.704973
67.5	91.28182268	6.5736833	+3.0473832	2.6555561	90.704844
90	91.15980423	5.2833659	+3.6768805	3.8659866	90.704445
112.5	91.02405557	3.8540430	+3.7764200	4.0781377	90.704075
135	90.89484135	2.5033169	+3.3308508	3.1725737	90.704038
157.5	90.79086206	1.4368221	+2.4080044	1.6581198	90.704442
180	90.72697664	0.8169216	+1.1483775	0.3771123	90.705145
202.5	90.71250894	0.7379931	−0.2562644	0.0187792	90.705802
225	90.75006384	1.2120495	−1.5920765	0.7248182	90.706059
247.5	90.83489465	2.1669211	−2.6556933	2.0167735	90.705781
270	90.95505790	3.4572383	−3.2851902	3.0861872	90.705088
292.5	91.09266180	4.8865615	−3.3847297	3.2760409	90.704321
315	91.22635509	6.2372881	−2.9391612	2.4702909	90.703855
337.5	91.33481374	7.3037834	−2.0163140	1.1625662	90.703855
Σ_1	728.48478638*	34.9624190‡	+1.5667604‡	14.9860379	725.637848
Σ_2	728.48478617	34.9624206	+1.5667608	14.9860312	725.637850

* $8a^2 + 4a^2e^2 + 8[a'^2 - 2kaa'ee' \cos K] = +728.48478640.$

† $8(a'^2e' - kaa'e \cos K] = +34.9624198.$

‡ $-8k'aa' \cos \varphi' \cdot e \sin K' = +1.5667610.$

ACTION OF SATURN ON MERCURY.

E	l	G	G'	G''	θ		
$0°$	$+0.41035\overline{1}$	90.704227	0.4147226	0.0043526	$3°$ $53'$ $50.78''$		
22.5	$+0.42248\overline{0}$	90.704715	0.4256040	0.0031100	3	56	31.53
45	$+0.380202$	90.704836	0.4105578	0.0302188	3	59	47.96
67.5	$+0.291021$	90.704520	0.3703889	0.0790439	4	2	4.89
90	$+0.16940\overline{3}$	90.703974	0.3081768	0.1383038	4	1	12.34
112.5	$+0.034023$	90.703579	0.2300012	0.1954823	3	55	22.98
135	$-0.09515\overline{4}$	90.703652	0.1455476	0.2403156	3	44	5.05
157.5	-0.199537	90.704241	0.0683029	0.2676388	3	29	2.13
180	-0.264126	90.705099	0.0149026	0.2789825	3	15	29.12
202.5	-0.279250	90.705800	0.0007394	0.2799874	3	11	3.15
225	$-0.24195\overline{3}$	90.705971	0.0294521	0.2713169	3	17	46.29
247.5	-0.156844	90.705536	0.0901199	0.2467190	3	29	20.24
270	$-0.03598\overline{7}$	90.704713	0.1675089	0.2031209	3	39	39.24
292.5	$+0.102383$	90.703922	0.2482641	0.1454820	3	46	28.97
315	$+0.236543$	90.703554	0.3215440	0.0847000	3	50	7.92
337.5	$+0.345001$	90.703713	0.3789646	0.0338216	3	52	2.72
Σ_1	$+0.559279$	725.636026	1.8124124	1.2513111	$29°$ $41'$ $58.70''$		
Σ_2	$+0.55927\overline{7}$	725.636026	1.8123850	1.2512850	29	41	56.61

ACTION OF SATURN ON MERCURY.

E	$\log K_0$	$\log L_0'$	$\log N_0$	$\log N$	$\log P$	$\log Q$
$0°$	0.00150875	0.27501245	0.17835367	5.6285747	$1.988290\overline{4}$	3.8492800
22.5	0.00154356	0.27505883	0.17840585	5.6455608	2.0053300	$3.866321\overline{9}$
45	0.00158664	0.27511623	0.17847042	$5.691930\overline{7}$	$2.051496\overline{7}$	$3.912626\overline{0}$
67.5	0.00161702	0.27515671	0.17851595	5.7568907	2.1160330	3.9773995
90	0.00160533	0.27514113	0.17849842	5.8276384	$2.186203\overline{5}$	4.0478488
112.5	0.00152867	0.27503898	0.17838353	5.8929411	2.2508613	4.1127652
135	0.00138529	0.27484791	0.17816860	$5.944592\overline{2}$	$2.301892\overline{3}$	$4.163986\overline{9}$
157.5	0.00120533	0.27460806	0.17789879	5.9773700	2.3341639	$4.196361\overline{7}$
180	0.00105402	0.27440638	0.17767193	5.9884810	2.3449566	4.2071875
202.5	0.00100673	0.27434335	0.17760102	5.9770716	$2.333467\overline{8}$	4.1956990
225	0.00107884	0.27443945	0.17770914	5.9440470	$2.300620\overline{5}$	4.1628231
247.5	0.00120882	0.27461270	0.17790402	$5.892240\overline{1}$	$2.249225\overline{8}$	$4.111330\overline{6}$
270	0.00133101	0.27477556	0.17808722	$5.826894\overline{2}$	$2.184467\overline{2}$	$4.046380\overline{1}$
292.5	0.00141515	0.27488768	0.17821333	$5.756216\overline{5}$	$2.114460\overline{0}$	3.9761077
315	0.00146116	0.27494901	0.17828232	5.6914235	2.0503133	3.9116762
337.5	0.00148559	0.27498156	0.17831893	5.6452895	2.0046971	3.8658214
Σ_1	0.01101104	1.19868812	1.42524172	$6.543581\overline{6}$	$7.408240\overline{3}$	$2.301808\overline{5}$
Σ_2	0.01101087	1.19868787	1.42524142	6.5435802	$7.408238\overline{8}$	$2.301806\overline{8}$

Action of Saturn on Mercury.

E	$\log V$	$J_1{}'$	J_2	J_3	F_2
0°	3.8492539	90.058195	−0.59752575	− 6.047237	+ 3.829727
22.5	3.8663033	89.638288	−0.20995430	− 7.910876	− 3.279413
45	3.9124452	89.669758	+0.36734212	− 8.328272	−10.040194
67.5	3.9769269	90.090673	+0.71855783	− 7.235732	−15.423345
90	4.0470220	90.582478	+0.67967125	− 4.799462	−18.609342
112.5	4.1115969	90.879839	+0.35660262	− 1.390299	−19.113128
135	4.1625507	90.892982	−0.04825504	+ 2.472681	−16.858027
157.5	4.1947623	90.670885	−0.36411899	+ 6.201244	−12.187337
180	4.2055204	90.332824	−0.49291773	+ 9.227617	− 5.812141
202.5	4.1940258	90.018885	−0.40939877	+11.090997	+ 1.296998
225	4.1612018	89.853321	−0.15646464	+11.507770	+ 8.057780
247.5	4.1098562̄	89.914463	+0.15541932	+10.414614	+13.440932
270	4.0451660	90.199819	+0.35977661	+ 7.978084	+16.626927
292.5	3.9752380	90.576579	+0.29271946	+ 4.569179	+17.130715
315	3.9111697	90.780941	−0.07365934	+ 0.706820	+14.875613
337.5	3.8656191	90.580690	−0.50178023	− 3.021123	+10.204923
Σ_1	2.2943297	722.370318*	+0.03796748	+12.718001	− 7.929657
Σ_2	2.2943284̄	722.370302	+0.03804694	+12.718004	− 7.929655

Action of Saturn on Mercury.

E	F_3	$10000 \times R_0$	$1000000 \times S_0$	$1000000 \times W_0$	$1000 \times R^{(n)}$	$1000000 \times S^{(n)}$
0°	−0.2217173	0.20835581	−0.38501122	− 4.2759258	0.0000000	−1.2520309
22.5	+0.0670688	0.21359630	−0.18752118	− 5.8140290	+2.6067657	−0.5980245
45	−0.2247552	0.23766012	+0.18723103	− 6.8102258	+5.0798256	+0.5659595
67.5	−0.9885705	0.27944181	+0.47990507̄	− 6.8742610	+7.2389467	+1.3456245
90	−1.8298399	0.33354940	+0.47167629	− 5.3763783	+8.6166490	+1.2184913
112.5	−2.2911663	0.39056810	+0.12052582	− 1.8384740	+8.6416600	−0.2886458
135	−2.1148434	0.43919690	−0.40799123	+ 3.5527836	+7.0044048	−0.9201906
157.5	−1.3919042	0.46950930	−0.83324243	+ 9.6804438	+3.9006063	−1.8089208
180	−0.5106652	0.47592790	−0.91983080	+14.8005740	0.0000000	−1.9709759
202.5	+0.0653940	0.45862525	−0.61203702	+17.3393050	−3.8101033	−1.3286969
225	+0.0610980	0.42282850	−0.06578321	+16.6811060	−6.7433594	−0.1483686
247.5	−0.4587079	0.37651300	+0.43874423	+13.4040390	−8.3306788	+1.0507430
270	−1.1366346	0.32772690	+0.65346841	+ 8.8350920	−8.4662365	+1.6881183
292.5	−1.5401525	0.28276850	+0.49946088	+ 4.2959022	−7.3251220	+1.4004581
315	−1.4203565	0.24587472	+0.10699355	+ 0.5601261	−5.2554073	+0.3234187
337.5	−0.8596732	0.22009701	−0.26508243	− 2.2258014	−2.6861019	−0.8453756
Σ_1	−7.3977141	2.69112025	−0.35924718	+27.9671518	+0.2358762	−0.4955782
Σ_2	−7.3977118	2.69111927	−0.35924707̄	+27.9671246	+0.2365727	−0.4955464

$\sin \varphi \cdot \tfrac{1}{2} A_1{}^{(s)} + \cos \varphi \cdot B_0{}^{(c)} = + 0.000000000034.$

* $\Sigma_1(J_1{}' − G'') = 721.119007.$ $\Sigma_2(J_1{}' − G'') = 721.119017.$

ACTION OF SATURN ON MERCURY.

E	$1000 \times [R_0 \sin v \\ + (\cos v + \cos E) S_0]$	$1000 \times \left[-R_0 \cos v \\ + \left(\frac{r}{a} \sec^2 \varphi + 1\right) \sin v \, S_0 \right]$	$1000000 \times W_0 \cos u$	$1000000 \times W_0 \sin u$	$1000 \times -2\frac{r}{a} R_0$
0°	−0.0007700023	−0.0208355810	− 3.7553310	− 2.0447571	−0.033103367
22.5	+0.009535644	−0.0190997850	− 3.2422806	− 4.8260289	−0.034604548
45	+0.019486084	−0.0136593860	− 0.8728046	− 6.7540641	−0.040621607
67.5	+0.027699067	.−0.0044469025	+ 2.0656152	− 6.5565758	−0.051490988
90	+0.032545329	+0.0078015052	+ 3.4868920	− 4.0923123	−0.066709877
112.5	+0.032625198	+0.0215155020	+ 1.6174933	− 0.8739008	−0.084259712
135	+0.027148276	+0.0344565700	− 3.5128323	+ 0.5313050	−0.100609845
157.5	+0.016337313	+0.0439769600	− 9.5267356	− 1.7182294	−0.111738865
180	+0.001839662	+0.0475927900	−12.9986018	− 7.0776661	−0.114756176
202.5	−0.013287669	+0.0439638930	−11.8447875	− 12.6630324	−0.109148595
225	−0.025446823	+0.0337808020	− 6.8547683	− 15.2076154	−0.096860250
247.5	−0.031966152	+0.0197522140	− 1.0475614	− 13.3630400	−0.081227519
270	−0.032206863	+0.0054309792	+ 2.5393222	− 8.4623118	−0.065545379
292.5	−0.027462505	−0.0063965152	.+ 2.7411608	− 3.3076901	−0.052103952
315	−0.019770536	−0.0145922640	+ 0.5055592	− 0.2411456	−0.042025673
337.5	−0.010655672	−0.0192899750	− 2.2254410	− 0.0400332	−0.035657730
Σ_1	+0.002825107	+0.0799753913	−21.4625646	−43.3484220	−0.560232174
Σ_2	+0.002825224	+0.0799754154	−21.4625365	−43.3485702	−0.560231909

DIFFERENTIAL COEFFICIENTS.

log coeff.

$[de/dt]_{00} = +\ 1.8596825 \ m' \quad p\ 0.2694389$

$[d\chi/dt]_{00} = +256.04618 \quad m' \quad p\ 2.4083183$

$[di/dt]_{00} = -\ 14.751452 \quad m' \quad n\ 1.1688348$

$[d\Omega/dt]_{00} = -244.39983 \quad m' \quad n\ 2.3881009$

$[d\pi/dt]_{00} = +254.22335 \quad m' \quad p\ 2.4052154$

$[dL/dt]_{00} = -373.17967 \quad m' \quad n\ 2.5719180$

FINAL VALUES CORRESPONDING TO THE ABOVE VALUE OF m'.

$[de/dt]_{00} = +0.00053109524$

$[d\chi/dt]_{00} = +0.073122627$

$[di/dt]_{00} = -0.0042127757$

$[d\Omega/dt]_{00} = -0.069796619$

$[d\pi/dt]_{00} = +0.072602050$

$[dL/dt]_{00} = -0.10657405$

COMPARISON WITH OTHER RESULTS.

	Leverrier.	Newcomb.	Method of Gauss.
$[de/dt]_{00}$	$+0.00053''$	$+0.00053''$	$+0.0005311''$
$e[d\pi/dt]_{00}$	$+0.01494$	$+0.01503$	$+0.0149273$
$[di/dt]_{00}$	-0.00421	-0.00421	-0.0042128
$\sin i \ [d\Omega/dt]_{00}$	-0.00853	-0.00850	-0.0085087
$[dL/dt]_{00}$	-0.1070		-0.106574

NOTES.

The considerable disagreement of the first sums is caused by the rather large value of e'. The very exact agreement toward the close of the computation shows, however, that all terms above the 15th order are wholly inappreciable, the total effect of all terms from the 8th to the 15th orders inclusive occurring with $[de/dt]_{00}$ and amounting to but 1/30000th of the value of this coefficient.

ACTION OF URANUS ON MERCURY.

E	A	$B \cos \epsilon$	$B \sin \epsilon$	g	h
0°	368.36907643	16.94656315	-5.8755608	27.995613	367.49553
45	368.87526360	22.1817430	-3.9726882	12.798540	367.49652
90	369.14240318	24.5977597	$+1.1481399$	1.069008	367.49606
135	369.01848670	22.7793354	$+6.4872119$	34.127748	367.49535
180	368.57162431	17.7916836	$+8.9169755$	64.480235	367.49625
225	368.05910248	12.5565017	$+7.0141016$	39.896574	367.49742
270	367.78562896	10.1404905	$+1.8932751$	2.906824	367.49671
315	367.91587927	11.9589113	-3.4457976	9.628773	367.49535
Σ_1	1473.86873228*	69.4764969†	$+6.0828297$‡	96.451680	1469.98455
Σ_2	1473.86873205	69.4764914	$+6.0828277$	96.451635	1469.98464

* $4a^2 + 2a^2e^2 + 4[a'^2 - 2kaa'ee' \cos K] = +1473.86873246.$

† $4[a'^2e' - kaa'e \cos K] = +69.4764933.$

‡ $-4k'aa' \cos \varphi' \cdot e \sin K' = +6.0828300.$

ACTION OF URANUS ON MERCURY.

E	l	G	G'	G''	θ ° ′ ″
0°	+0.06261	367.49532	0.30919	0.24638	2 13 39.23
45	+0.56779	367.49643	0.6237$\overline{3}$	0.0558$\overline{4}$	2 27 51.84
90	+0.83539	367.49605	0.8388$\overline{8}$	0.00347	2 44 38.85
135	+0.71220	367.49510	0.8250$\overline{1}$	0.1125$\overline{7}$	2 53 41.24
180	+0.26443	367.49577	0.5717$\overline{8}$	0.3068$\overline{7}$	2 48 5.47
225	−0.24927	367.49712	0.22774	0.47670	2 30 27.70
270	−0.52202	367.49669	0.0147$\overline{4}$	0.53674	2 13 6.46
315	−0.39041	367.49528	0.05839	0.44873	2 7 39.31
Σ_1	+0.64041	1469.98383	1.7345$\overline{8}$	1.0934$\overline{6}$	9 59 30.01
Σ_2	+0.64031	1469.98393	1.73486	1.09383	9 59 40.09

E	$\log K_0$	$\log L_0'$	$\log N_0$	$\log N$	$\log P$	$\log Q$
0°	0.00049251	0.27365789	0.17682994	4.7157168	9.8582891	2.3270039
45	0.00060285	0.27380499	0.17699541	4.7796309	9.9227977	2.3913072
90	0.00074754	0.27399787	0.17721240	4.916327$\overline{6}$	0.059812$\overline{1}$	2.528283$\overline{3}$
135	0.00083193	0.27411036	0.17733893	5.0341227	0.1774639	2.6460770
180	0.00077916	0.27404002	0.17725980	5.0782321	0.2210425	2.6898771
225	0.00062422	0.27383348	0.17702747	5.033266$\overline{9}$	0.175467$\overline{0}$	2.644477$\overline{7}$
270	0.00048849	0.27365254	0.17682391	4.915122$\overline{7}$	0.057000$\overline{6}$	2.526059$\overline{3}$
315	0.00044928	0.27360025	0.17676511	4.778783$\overline{5}$	9.9208203	2.380766$\overline{0}$
Σ_1	0.00250770	1.09534833	0.70812605	9.6253991	0.1961442	0.0712235
Σ_2	0.00250828	1.09534908	0.70812692	9.6258039	0.196548$\overline{9}$	0.0716217

E	$\log V$	J_1'	J_2	J_3	F_2
0°	2.3266400	366.509888	−2.2679092	−1.6777615	+100.81429
45	2.3912247	363.119586	−0.5138777	−3.4254958	+ 68.16435
90	2.528278$\overline{2}$	365.160838	+2.2768225	−2.9100648	− 19.70006
135	2.6459108	367.567473	+0.7238013	−0.4333073	−111.30915
180	2.6894241	366.570373	−1.5747120	+2.5538323	−152.99965
225	2.6437740	364.094438	−1.1669936	+4.3014333	−120.34967
270	2.5252669	364.090394	+1.5139175	+3.7858697	− 32.48527
315	2.3891044	367.299619	+1.4067899	+1.3092445	+ 59.12384
Σ_1	0.069609$\overline{2}$	1462.331493*	−0.0518812	+1.7518757	−104.37069
Σ_2	0.0700139	1462.082117	−0.4497199	+1.7518747	−104.37063

* $\Sigma_1(J_1' - G'') = 1461.238038.$

$\Sigma_2(J_1' - G'') = 1460.987287.$

ACTION OF URANUS ON MERCURY.

E	F_2	$1000000 \times R_0$	$1000000000 \times S_0$	$1000000 \times W_0$	$100000 \times R^{(n)}$	$1000000 \times S^{(n)}$
0°	− 7.546840	2.572319	− 40.838728	−0.3564792	0.00000000	−0.13280483
45	− 0.473969	2.902866	− 6.943520	−0.8432739	+0.62046843	−0.02098878
90	− 1.504590	4.048391	+ 74.582644	−0.9823294	+1.04582909	+0.19267091
135	−14.059822	5.410853	+ 15.278801	−0.1938530	+0.86293520	+0.03446006
180	−17.382080	5.928252	−102.476485	+1.2462639	0.00000000	−0.21958237
225	− 5.500356	5.225873	− 69.412255	+1.8932075	−0.83343288	−0.15655363
270	+ 1.222261	3.978026	+ 47.038051	+1.2690546	−1.02765154	+0.12151438
315	− 5.177088	2.987389	+ 39.388463	+0.3202880	−0.63853471	+0.11906293
Σ_1	−25.211249	16.526988	− 21.694518	+1.1765099	+0.01817755	−0.03820191
Σ_2	−25.211235	16.526981	− 21.688511	+1.1763686	+0.01143604	−0.02401942

E	$1000000 \times [R_0 \sin v + (\cos v + \cos E)S_0]$	$1000000 \times \left[-R_0 \cos v + \left(\frac{r}{a}\sec^2\varphi+1\right)\sin v S_0 \right]$	$1000000 \times W_0 \cos u$	$1000000 \times W_0 \sin u$	$1000000 \times -2\frac{r}{a}R_0$
0°	−0.0816775	−2.5723195	−0.31307783	−0.17046914	− 4.0868774
45	+2.3415248	−1.7140875	−0.10807474	−0.83631981	− 4.9616693
90	+3.9465628	+0.9815683	+0.63709735	−0.74771500	− 8.0967815
135	+3.2460626	+4.3319665	+0.19167313	−0.02898997	−12.3950199
180	+0.2049530	+5.9282527	−1.09453081	−0.59596603	−14.2942599
225	−3.0528880	+4.2563802	−0.77797582	−1.72597460	−11.9712672
270	−3.9027069	+0.7238029	+0.36474303	−1.21550896	− 7.9560519
315	−2.3679844	−1.8134015	+0.28908589	−0.13789041	− 5.1061400
Σ_1	+0.1671314	+5.0613044	−0.40576826	−2.72965913	−34.4339707
Σ_2	+0.1667150	+5.0608577	−0.40529154	−2.72917479	−34.4340964

$\sin \varphi \cdot \tfrac{1}{2}A_1^{(s)} + \cos \varphi \cdot B_0^{(c)} = -0.00000000000024.$

DIFFERENTIAL COEFFICIENTS.

log coeff.

$$[de/dt]_{00} = + 0.21975650'' \; m' \quad p \; 9.3419417$$
$$[d\chi/dt]_{00} = +32.406731 \; m' \quad p \; 1.5106352$$
$$[di/dt]_{00} = - 0.55745051 \; m' \quad n \; 9.7462063$$
$$[d\Omega/dt]_{00} = -30.777028 \; m' \quad n \; 1.4882267$$
$$[d\pi/dt]_{00} = +32.177180 \; m' \quad p \; 1.5075480$$
$$[dL/dt]_{00} = -45.859693 \; m' \quad n \; 1.6614312$$

FINAL VALUES CORRESPONDING TO THE ABOVE VALUE OF m'.

$$[de/dt]_{00} = +0.\overset{''}{0}000096384435$$
$$[d\chi/dt]_{00} = +0.0014213479$$
$$[di/dt]_{00} = -0.000024449584$$
$$[d\Omega/dt]_{00} = -0.0013498699$$
$$[d\pi/dt]_{00} = +0.0014112801$$
$$[dL/dt]_{00} = -0.0020113907$$

COMPARISON WITH OTHER RESULTS.

	Leverrier.	Newcomb.	Method of Gauss.
$[de/dt]_{00}$	$+0.00000$	$+0.00001$	$+0.0000096$
$e[d\pi/dt]_{00}$	$+0.00029$	$+0.00029$	$+0.0002902$
$[di/dt]_{00}$	-0.00001	-0.00002	-0.0000244
$\sin i\ [d\Omega/dt]_{00}$	-0.00016	-0.00016	-0.0001646

NOTES.

In the results of this computation, published in *A. J.*, No. 398, the residual from the test equation which arises from the constancy of the major axis was stated very much too large. Its true value is as here given.

A comparison of the above figures with the corresponding tabulation for Saturn on Mercury and Uranus on Venus shows that a division into eight parts is fully sufficient. The effect of all terms of the 4th and higher orders may, however, in some cases amount to 1/1000th of the whole.

ACTION OF NEPTUNE ON MERCURY.

E	A	$B \cos \epsilon$	$B \sin \epsilon$	g	h
$0°$	904.45979027	$+15.1625927$	-5.380877	1.889897	904.17356
45	904.50658592	$+17.0408109$	$+3.150684$	0.647955	904.17446
90	904.46648308	$+12.3030460$	$+10.652425$	7.406776	904.17363
135	904.36745259	$+3.7246157$	$+12.729927$	10.577523	904.17288
180	904.26302599	-3.6693551	$+8.166225$	4.352861	904.17435
225	904.20989585	-5.5475736	-0.365336	0.008712	904.17570
270	904.24366422	-0.8098086	-7.867076	4.039717	904.17446
315	904.34902898	$+7.7686225$	-9.944584	6.455136	904.17292
Σ_1	3617.43296356*	$+22.9864685$†	$+5.570698$‡	17.689251	3616.69599
Σ_2	3617.43296334	$+22.9864755$	$+5.570691$	17.689326	3616.69596

* $4a^2 + 2a^2e^2 + 4[a'^2 - 2kaa'ee' \cos K] = +3617.4329635.$

† $4[a'^2e' - kaa'e \cos K] = +22.986469.$

‡ $-4k'aa' \cos \varphi' \cdot e \sin K' = +5.570695.$

THE SECULAR VARIATIONS OF THE ELEMENTS

ACTION OF NEPTUNE ON MERCURY.

E	l	G	G'	G''	θ		
					°	′	″
0°	0.22096	904.17356	0.2300460	0.0090860	0	55	54.56
45	0.26686	904.17446	0.2695190	0.0026589	0	59	38.88
90	$0.22759\overline{9}$	904.17362	0.2591945	0.0316047	1	1	39.23
135	$0.12931\overline{1}$	904.17286	0.1906740	0.0613537	0	57	23.73
180	0.02341	904.17434	0.0820755	0.0586556	0	42	53.31
225	0.03107	904.17570	0.0003070	0.0313770	0	20	20.99
270	$0.00394\overline{4}$	904.17445	0.0688415	0.0649018	0	41	48.59
315	0.11084	904.17291	0.1564755	0.0456255	0	51	23.82
Σ_1	0.57589	3616.69597	0.6401575	0.1642481	3	22	15.69
Σ_2	$0.57594\overline{4}$	3616.69593	0.6169755	0.1410151	3	8	47.42

E	$\log K_0$	$\log L_0'$	$\log N_0$	$\log N$	$\log P$	$\log Q$
0°	0.00008616	0.27311614	0.17622049	4.1292404	8.4898441	1.3492047
45	0.00009807	0.27313202	0.17623836	$4.1927246\overline{6}$	$8.5533496\overline{6}$	$1.4127095\overline{5}$
90	0.00010478	0.27314097	0.17624841	4.3291694	8.6897764	1.5491508
135	0.00009080	0.27312233	0.17622746	$4.4470372\overline{2}$	$8.8075977\overline{7}$	$1.6669838\overline{8}$
180	0.00005070	0.27306887	0.17616731	$4.4915054\overline{4}$	$8.8520137\overline{7}$	$1.7113924\overline{4}$
225	0.00001141	0.27301649	0.17610837	$4.4469773\overline{3}$	$8.8074580\overline{0}$	$1.6668178\overline{8}$
270	0.00004818	0.27306552	0.17616353	4.3290882	8.6895869	1.5489683
315	0.00007281	0.27309835	0.17620048	4.1926694	$8.5532210\overline{0}$	1.4125965
Σ_1	0.00028982	1.09239150	0.70479974	$7.2790034\overline{4}$	$4.7212211\overline{1}$	$6.1587162\overline{2}$
Σ_2	0.00027309	1.09236919	0.70477467	$7.2794084\overline{4}$	4.7216261	$6.1591075\overline{5}$

E	$\log V$	J_1'	J_2	J_3	F_2
0°	1.3491992	897.83542	−6.8010114	− 59.717292	+41.026915
45	1.4127079	890.85490	+1.7283795	− 93.077013	−24.022644
90	1.5491318	899.76750	+6.4446186	− 62.851349	−81.220264
135	1.6669469	904.08417	−1.3097917	+ 13.254268	−97.060333
180	1.7113571	897.88500	−6.6871696	+ 90.657875	−62.264029
225	1.6667989	891.04449	−2.2340656	+124.017084	+ 2.785531
270	1.5489293	894.36354	+5.9480179	+ 93.790933	+59.983153
315	1.4125691	903.73833	+2.4206062	+ 17.685807	+75.823263
Σ_1	6.1586174	3589.85146*	−1.0955445	+ 61.880167	−42.474225
Σ_2	6.1590228	3589.72189	+0.6051284	+ 61.880146	, −42.474183

* $\Sigma_1(J_1' - G'') = 3589.68721.$

$\Sigma_2(J_1' - G'') = 3589.58087.$

ACTION OF NEPTUNE ON MERCURY.

E	F_3	$1000000 \times R_0$	$1000000000 \times S_0$	$1000000000 \times W_0$	$1000000 \times R^{(n)}$	$1000000 \times S^{(n)}$
0°	− 2.928627	0.6591808	−15.070783	−133.45316	0.000000	−0.04900917
45	− 0.320722	0.7449111	+ 4.384511	−240.74232	+1.592198	+0.01325344
90	− 8.201274	1.0513250	+22.423107	−222.59983	+2.715911	+0.05792607
135	−13.630154	1.3989891	− 6.706655	+ 61.47306	+2.231138	−0.01512625
180	− 6.745208	1.5178690	−34.846090	+466.35639	0.000000	−0.07466679
225	+ 0.065738	1.3382857	−10.354891	+575.81202	−2.134326	−0.02335460
270	− 3.847073	1.0317860	+21.345904	+331.94457	−2.665434	+0.05514330
315	− 7.846125	0.7779772	+ 6.529864	+ 45.70113	−1.662874	+0.01973839
Σ_1	−21.722182	4.2601608	− 6.147862	+442.24897	+0.050477	−0.01060659
Σ_2	−21.731263	4.2601631	− 6.147171	+442.24389	+0.026136	−0.00548902

E	$1000000000 \times [R_0 \sin v + (\cos v + \cos E) S_0]$	$1000000000 \times \left[-R_0 \cos v + \left(\frac{r}{a} \sec^2 \varphi + 1\right) \sin v S_0\right]$	$1000000000 \times W_0 \cos u$	$1000000000 \times W_0 \sin u$	$1000000000 \times -2 \frac{r}{a} R_0$
0°	− 30.1416	− 659.1808	−117.20523	− 63.81760	−1047.3000
45	+ 608.8428	− 430.4075	− 30.85375	−238.75703	−1273.2255
90	+1024.2533	+ 261.0141	+144.36884	−169.43527	−2102.6500
135	+ 855.3050	+1105.9013	− 60.78178	+ 9.19306	−3204.7615
180	+ 69.6922	+1517.8690	−409.57745	−223.01267	−3659.9008
225	− 792.9698	+1080.1647	−236.61842	−524.94878	−3065.7028
270	−1014.1308	+ 169.4383	+ 95.40526	−317.93876	−2063.5719
315	− 621.4949	− 466.5348	+ 41.24897	− 19.67525	−1329.7429
Σ_1	+ 49.6731	+1289.1406	−287.00858	−774.20430	−8873.4227
Σ_2	+ 49.6831	+1289.1237	−287.00498	−774.18800	−8873.4327

$\sin \varphi \cdot \frac{1}{2} A_1^{(s)} + \cos \varphi \cdot B_0^{(c)} = + 0.000000000000013.$

DIFFERENTIAL COEFFICIENTS.

				log coeff.
$[de/dt]_{00}$	=	+ 0.065401848″	m'	p 8.8155900
$[d\chi/dt]_{00}$	=	+ 8.2544736	m'	p 0.9166894
$[di/dt]_{00}$	=	− 0.39452600	m'	n 9.5960756
$[d\Omega/dt]_{00}$	=	− 8.7298700	m'	n 0.9410078
$[d\pi/dt]_{00}$	=	+ 8.1893623	m'	p 0.9132501
$[dL/dt]_{00}$	=	−11.826130	m'	n 1.0728427

FINAL VALUES CORRESPONDING TO THE ABOVE VALUE OF m'.

$$[de/dt]_{00} = +0.\overset{''}{0}0000331989$$
$$[d\chi/dt]_{00} = +0.00041900885$$
$$[di/dt]_{00} = -0.00002002670$$
$$[d\Omega/dt]_{00} = -0.00044314061$$
$$[d\pi/dt]_{00} = +0.00041570371$$
$$[dL/dt]_{00} = -0.00060031125$$

COMPARISON WITH OTHER RESULTS.

	Leverrier.	Newcomb.	Method of Gauss.
$[de/dt]_{00}$	$+0.\overset{''}{0}0000$	$+0.\overset{''}{0}0000$	$+0.\overset{''}{0}000033$
$e[d\pi/dt]_{00}$	$+0.00009$	$+0.00009$	$+0.0000855$
$[di/dt]_{00}$	-0.00001	-0.00002	-0.0000200
$\sin i\ [d\Omega/dt]_{00}$	-0.00005	-0.00005	-0.0000508

NOTES.

In the final results of this computation, published in *A. J.*, No. 398, the value of the residual arising from the equation $[da/dt]_{00} = 0$ is greatly overstated. Its true value is that given above.

The very large disagreement in G', G'', θ, etc., arises from the large values of e' and I but the gradual lessening of the discrepancies as the end of the computation is approached shows that terms of the 8th and higher orders are wholly inappreciable. The greatest effect produced by all terms of the 4th and higher orders here occurs with $[de/dt]_{00}$ and amounts to but 1/10000th of the value of this coefficient.

VENUS.

ACTION OF MERCURY ON VENUS.

E	A	$B \cos \epsilon$	$B \sin \epsilon$	$1000 \times g$	h
0°	0.73249627	+0.19271542	+0.22036223	0.30759837	0.58840054
15	0.70628935	+0.12839518	+0.25427819	0.40956972	0.56402284
30	0.67778527	+0.05734848	+0.27076206	0.46439247	0.53673295
45	0.64892716	−0.01558282	+0.26869056	0.45731389	0.50812387
60	0.62168257	−0.08542863	+0.24820463	0.39023784	0.48000449
75	0.59791027	−0.14742905	+0.21070058	0.28121662	0.45436878
90	0.57923027	−0.19735886	+0.15873412	0.15960636	0.43333042
105	0.56691636	−0.23181556	+0.09584678	0.05819218	0.41895978
120	0.56180731	−0.24845074	+0.02632410	0.00438952	0.41292353
135	0.56425026	−0.24613080	−0.04509594	0.01288203	·0.41598961
150	0.57407787	−0.22501408	−0.11354628	0.08166866	0.42773114
165	0.59061890	−0.18653936	−0.17436209	0.19258116	0.44670373̄
180	0.61274521	−0.13332873	−0.22339892	0.31613445	0.47092914
195	0.63894853	−0.06900848	−0.25731488	0.41941058	0.49830819
210	0.66744400	+0.00203821	−0.27379880	0.47486769	0.52683118
225	0.69628968	+0.07496950	−0.27172719	0.46770903	0.55465379
240	0.72352217	+0.14481528	−0.25124139	0.39984532	0.58013161
255	0.74728586	+0.20681571	−0.21373734	0.28938127	0.60184004̄
270	0.76596211	+0.25674559	−0.16177082	0.16577156	0.61860745
285	0.77827946	+0.29120222	−0.09888350	0.06193800	0.62954599̄
300	0.78339732	+0.30783739	−0.02936080	0.00546066	0.63407808
315	0.78096661	+0.30551754	+0.04205926	0.01120554	0.63196045
330	0.77115125	+0.28440072	+0.11050957	0.07735875	0.62329194
345	0.75461912	+0.24592608	+0.17132542	0.18593159	0.60851491
Σ_1	8.07130162*	0.35632005†	−0.01822030‡	2.84733165	6.33299247
Σ_2	8.07130156	0.35632016	−0.01822015	2.84733161	6.33299197

* $12a^2 + 6a^2e^2 + 12[a'^2 - 2kaa'ee' \cos K] = 8.07130158.$

† $12[a'^2e' - kaa'e \cos K] = + 0.35632010.$

‡ $- 12k'aa' \cos \varphi' \cdot e \sin K' = - 0.01822024.$

ACTION OF MERCURY ON VENUS.

E	l	G	G'	G''	ʻ		
°					°	′	″
0	0.13776127	0.58723527	0.14259981	0.00367327	29	50	12.97
15	0.13593205	0.56231468	0.14274284	0.00510263	30	41	37.56
30	0.13471786	0.53456025	0.14296704	0.00607649	31	40	18.70
45	0.13446883	0.50568767	0.14321940	0.00631437	32	42	44.77
60	0.13534362	0.47761736	0.14342737	0.00569662	33	44	34.65
75	0.13720703	0.45239646	0.14351084	0.00433149	34	40	36.18
90	0.13956539	0.43206750	0.14340426	0.00257595	35	25	4.49
105	0.14162212	0.41845750	0.14309623	0.00097183	35	52	44.73
120	0.14254931	0.41288415	0.14266320	0.00007451	36	0	33.62
135	0.14192619	0.41587654	0.14225701	0.00021775	35	48	50.91
150	0.14001227	0.42706500	0.14202490	0.00134648	35	20	41.47
165	0.13758072	0.44529837	0.14203102	0.00304495	34	40	10.49
180	0.13548161	0.46890720	0.14224327	0.00473972	33	51	11.13
195	0.13430588	0.49597000	0.14257523	0.00593116	32	57	11.97
210	0.13427836	0.52451120	0.14293246	0.00633413	32	1	25.39
225	0.13530143	0.55262582	0.14323803	0.00590863	31	6	52.56
240	0.13705610	0.57856627	0.14343948	0.00481804	30	16	21.83
255	0.13911137	0.60079677	0.14351100	0.00335627	29	32	27.98
270	0.14102020	0.61804514	0.14345226	0.00186975	28	57	29.65
285	0.14239902	0.62934391	0.14328793	0.00068684	28	33	26.40
300	0.14298478	0.63406059	0.14306246	0.00006020	28	21	52.79
315	0.14267170	0.63192420	0.14283210	0.00012415	28	23	50.37
330	0.14152485	0.62303406	0.14265313	0.00087040	28	39	39.46
345	0.13976975	0.60786141	0.14256873	0.00214548	29	8	52.62
Σ_1	1.66229562	6.31855399	1.71486964	0.03813556	384	9	26.15
Σ_2	1.66229608	6.31855323	1.71487036	0.03813555	384	9	26.54

ACTION OF MERCURY ON VENUS.

E	$\log K_0$	$\log L_0'$	$\log N_0$	$\log N$	$\log P$	$\log Q$
0°	0.09428087	0.39679023	0.31472571	0.0090498	0.8627996	0.5522553
15	0.10017341	0.40440450	0.32320948	0.0415726	0.9381720	0.610879$\bar{6}$
30	0.10716555	0.41342180	0.33324961	0.0806578	1.0282684	0.6810018
45	0.11492344	0.42340404	0.34435531	0.124816$\bar{3}$	1.129676$\bar{9}$	0.759899$\bar{9}$
60	0.12293987	0.43369427	0.35579389	0.171632$\bar{4}$	1.2368681	0.843197$\bar{0}$
75	0.13049856	0.44337414	0.36654485	0.2174860	1.3415449	0.9243732
90	0.13670353	0.45130400	0.37534554	0.2575171	1.4325547	0.9947294
105	0.14065788	0.45634998	0.38094249	0.2862203	1.4972527	1.0445040
120	0.14178785	0.45779080	0.38254019	0.2989083	1.5248859	1.0655419
135	0.14009658	0.45563407	0.38014861	0.2935151	1.5107656	1.0544719
150	0.13608370	0.45051253	0.37446742	0.271437$\bar{9}$	1.458228$\bar{2}$	1.014044$\bar{2}$
165	0.13043971	0.44329885	0.36646127	0.236759$\bar{5}$	1.3768369	0.9516100
180	0.12381666	0.43481823	0.35704266	0.194571$\bar{5}$	1.2784801	0.8761594
195	0.11676724	0.42577304	0.34698954	0.1495758	1.1741125	0.795947$\bar{2}$
210	0.10975153	0.41675185	0.33695547	0.105445$\bar{2}$	1.072260$\bar{8}$	0.7174326
225	0.10314673	0.40824135	0.32748247	0.064777$\bar{9}$	0.9789193	0.6452104
240	0.09725105	0.40063000	0.31900464	0.029299$\bar{4}$	0.898020$\bar{0}$	0.582349$\bar{3}$
255	0.09229597	0.39422222	0.31186325	0.000126$\bar{3}$	0.8320547	0.530842$\bar{6}$
270	0.08845847	0.38925289	0.30632243	9.977973$\bar{8}$	0.782562$\bar{5}$	0.491964$\bar{1}$
285	0.08587486	0.38590395	0.30258708	9.963306$\bar{2}$	0.7504867	0.4665316
300	0.08464926	0.38431433	0.30081370	9.9564277	0.7363980	0.455069$\bar{4}$
315	0.08485629	0.38458290	0.30111334	9.9575313	0.7406137	0.457894$\bar{4}$
330	0.08653835	0.38676424	0.30354673	9.966711$\bar{9}$	0.7632399	0.475140$\bar{5}$
345	0.08969689	0.39085723	0.30811150	9.9839477	0.8041353	0.5067244
Σ_1	1.32942669	5.01604517	4.05980799	1.3196325	13.074566$\bar{1}$	8.748884$\bar{7}$
Σ_2	1.32942756	5.01604627	4.05980919	1.3196348	13.0745711	8.748889$\bar{0}$

ACTION OF MERCURY ON VENUS.

E	$\log V$	J_1'	J_2	J_3	$1000 \times F_2$
0°	0.5489818	0.14641488	+0.046569366	+0.007078444	−6.6249758
15	0.6061455	0.14799570	+0.053805593	+0.005998321	−7.6446250
30	0.6750909	0.14917691	+0.057280924	+0.004274648	−8.1401963
45	0.7534266	0.14962155	+0.056746961	+0.002024896	−8.0779189
60	0.8370288	0.14915409	+0.052252409	−0.000597633	−7.4620305
75	0.9194294	0.14784239	+0.044134510	−0.003414224	−6.3345072
90	0.9916527	0.14602955	+0.032988273	−0.006232935	−4.7721868
105	1.0433053	0.14427332	+0.019614339	−0.008861684	−2.8815397
120	1.0654487	0.14317010	+0.004953440	−0.011121317	−0.7914084
135	1.0542014	0.14310862	−0.009985226	−0.012857840	+1.3557653
150	1.0124149	0.14408811	−0.024195183	−0.013952904	+3.4136578
165	0.9480753	0.14573205	−0.036737898	−0.014331875	+5.2420265
180	0.8709298	0.14748132	−0.046798849	−0.013968920	+6.7162708
195	0.7897501	0.14882171	−0.053732494	−0.012888777	+7.7359211
210	0.7111602	0.14942892	−0.057095855	−0.011165046	+8.2314924
225	0.6396416	0.14920890	−0.056671026	−0.008915214	+8.1692132
240	0.5779992	0.14827058	−0.052477157	−0.006292610	+7.5533276
255	0.5279163	0.14686716	−0.044770588	−0.003475962	+6.4258044
270	0.4903757	0.14532896	−0.034036602	−0.000657227	+4.8634820
285	0.4659576	0.14399775	−0.020965237	+0.001971497	+2.9728356
300	0.4550194	0.14316554	−0.006416142	+0.004231074	+0.8827039
315	0.4577910	0.14302191	+0.008631431	+0.005967519	−1.2644706
330	0.4744060	0.14361483	+0.023141851	+0.007062502	−3.3223626
345	0.5048724	0.14483260	+0.036096092	+0.007441419	−5.1507310
Σ_1	8.7105081	1.75532379*	−0.003833505	−0.041341924	+0.5477741
Σ_2	8.7105125	1.75532376	−0.003833543	−0.041341924	+0.5477737

$* \; \Sigma_1(J_1' - G'') = 1.71718823.$

$\Sigma_2(J_2' - G'') = 1.71718811.$

ACTION OF MERCURY ON VENUS.

E	$1000 \times F_3$	R_0	S_0	W_0	$R^{(n)}$	$S^{(n)}$
0°	+0.12855762	−1.0113711	+0.11654330	+0.02599380	0.0000000	+0.16223015
15	+0.28853773	−1.0855630	+0.15095431	+0.02672256	−0.3910151	+0.21008155
30	+0.43551240	−1.1839703	+0.18420453	+0.02487781	−0.8232930	+0.25617923
45	+0.52994293	−1.3088599	+0.21275230	+0.01862044	−1.2857216	+0.29555816
60	+0.54641763	−1.4594255	+0.23029016	+0.00532100	−1.7533282	+0.31946706
75	+0.48046967	−1.6276742	+0.22753508	−0.01781201	−2.1774255	+0.31512321
90	+0.34977540	−1.7935752	+0.19439994	−0.05167293	−2.4796011	+0.26875615
105	+0.18941983	−1.9243133	+0.12616235	−0.09195653	−2.5651527	+0.17410988
120	+0.04248891	−1.9847155	+0.03108865	−0.12787906	−2.3681413	+0.04283321
135	−0.05148113	−1.9570464	−0.06917608	−0.14733858	−1.9039342	−0.09517474
150	−0.06711060	−1.8525243	−0.15091802	−0.14550282	−1.2730044	−0.20741354
165	+0.00001044	−1.7023416	−0.20114407	−0.12716787	−0.6051233	−0.27625376
180	+0.13212520	−1.5390947	−0.22013939	−0.10126618	0.0000000	−0.30227222
195	+0.29405223	−1.3852923	−0.21561000	−0.07503510	+0.4924233	−0.29612150
210	+0.44259806	−1.2520945	−0.19638955	−0.05218717	+0.8604054	−0.26990714
225	+0.53811675	−1.1428753	−0.16935246	−0.03375802	+1.1118587	−0.23300075
240	+0.55512295	−1.0569737	−0.13887026	−0.01942443	+1.2611694	−0.19133211
255	+0.48911303	−0.9922445	−0.10732663	−0.00839923	+1.3226839	−0.14811570
270	+0.35776744	−0.9463685	−0.07579537	+0.00013574	+1.3083458	−0.10478640
285	+0.19621617	−0.9174587	−0.04456334	+0.00686904	+1.2273328	−0.06171770
300	+0.04762640	−0.9042927	−0.01348260	+0.01232297	+1.0864013	−0.01870356
315	−0.04835263	−0.9063942	+0.01780841	+0.01685710	+0.8903706	+0.02473967
330	−0.06620421	−0.9240677	+0.04973149	+0.02067163	+0.6425655	+0.06916321
345	−0.00136711	−0.9584216	+0.08262385	+0.02378861	+0.3452194	+0.11498675
Σ_1	+2.90467772	−5.9084737	+0.01066288	−0.40860954	−3.5384806	+0.02421404
Σ_2	+2.90467759	−5.9084840	+0.01066372	−0.40860959	−3.5384837	+0.02421607

$\sin \varphi \cdot \tfrac{1}{2} A_1^{(s)} + \cos \varphi \cdot B_0^{(c)} = + 0.000000033.$

ACTION OF MERCURY ON VENUS.

E	$[R_0 \sin v + (\cos v + \cos E)S_0]$	$\left[- R_0 \cos v + \left(\frac{r}{a}\sec^2\varphi + 1\right)\sin v S_0\right]$	$W_0 \cos u$	$W_0 \sin u$	$-2\frac{r}{a}R_0$
0°	+0.2330866	+1.0113711	+0.015230839	+0.021064164	2.0089005
15	+0.0087244	+1.1264720	+0.009475283	+0.024986282	2.1567748
30	−0.2767659	+1.2080644	+0.002459142	+0.024755972	2.3539076
45	−0.6298334	+1.2226121	−0.003043897	+0.018369962	2.6050539
60	−1.0391050	+1.1217558	−0.002204101	+0.004843036	2.9088638
75	−1.4586399	+0.8508163	+0.011332120	−0.013742297	3.2495836
90	−1.7948636	+0.3765263	+0.042079553	−0.029990049	3.5871504
105	−1.9215248	−0.2668018	+0.086138100	−0.032190511	3.8554443
120	−1.7441606	−0.9487539	+0.127283754	−0.012324508	3.9830138
135	−1.2790794	−1.4880990	+0.145357191	+0.024082111	3.9330324
150	−0.6591294	−1.7579567	+0.132584634	+0.059936528	3.7270060
165	−0.0490234	−1.7488888	+0.098503636	+0.080428204	3.4271881
180	+0.4402788	−1.5390947	+0.059336027	+0.082061396	3.0992536
195	+0.7728021	−1.2274791	+0.026854296	+0.070065081	2.7888981
210	+0.9628351	−0.8906639	+0.005513895	+0.051895060	2.5190296
225	+1.0443017	−0.5731024	−0.005195888	+0.033355748	2.2968106
240	+1.0518039	−0.2937729	−0.007836004	+0.017773731	2.1211810
255	+1.0129579	−0.0559799	−0.005257515	−0.006550236	1.9880041
270	+0.9468650	+0.1451147	+0.000109448	−0.000080285	1.8927371
285	+0.8649660	+0.3177537	+0.006402057	−0.002489441	1.8316679
300	+0.7723975	+0.4708818	+0.012250672	−0.001332937	1.8023971
315	+0.6691423	+0.6125550	+0.016656295	+0.002594179	1.8040170
330	+0.5508294	+0.7487960	+0.018894170	+0.008386104	1.8371831
345	+0.4092814	+0.8824106	+0.018479728	+0.014979907	1.9041728
Σ_1	−0.5559282	−0.3477320	+0.405702029	+0.226988212	31.8406236
Σ_2	−0.5559251	−0.3477313	+0.405701406	+0.226989461	31.8406376

DIFFERENTIAL COEFFICIENTS.

log coeff.

$$[de/dt]_{00} = - \quad 97592.111 \ m' \quad n \ 4.9894147$$
$$[d\chi/dt]_{00} = -8920493.9 \quad m' \quad n \ 6.9503889$$
$$[di/dt]_{00} = + \quad 71223.820 \ m' \quad p \ 4.8526253$$
$$[d\Omega/dt]_{00} = + \quad 673299.06 \ m' \quad p \ 5.8282080$$
$$[d\pi/dt]_{00} = -8919313.6 \quad m' \quad n \ 6.9503315$$
$$[dL/dt]_{00} = +5590689.3 \quad m' \quad p \ 6.7474654$$

FINAL VALUES CORRESPONDING TO THE ABOVE VALUE OF m'.

$$[de/dt]_{00} = -0.\overset{''}{0}13012279$$
$$[d\chi/dt]_{00} = -1.1893992$$
$$[di/dt]_{00} = +0.0094965089$$
$$[d\Omega/dt]_{00} = +0.089773204$$
$$[d\pi/dt]_{00} = -1.1892420$$
$$[dL/dt]_{00} = +0.74542525$$

COMPARISON WITH OTHER RESULTS.

	Leverrier.	Newcomb.	Method of Gauss.
$[de/dt]_{00}$	$-0.\overset{''}{0}1304$	$-0.\overset{''}{0}1301$	$-0.\overset{''}{0}13012$
$e[d\pi/dt]_{00}$	-0.00810	-0.00814	-0.008138
$[di/dt]_{00}$	$+0.00950$	$+0.00949$	$+0.009497$
$\sin i \; [d\Omega/dt]_{00}$	$+0.00529$	$+0.00531$	$+0.005301$
$[dL/dt]_{00}$	$+0.747$		$+0.745425$

NOTES.

This computation was originally made with but twelve points of division, but it was found that, notwithstanding the small eccentricity of the orbit of Venus, the values of e' and I are here so large that the tests which arise by comparing the sums of the functions were, toward the close of the computation, entirely inapplicable. The sums for $[de/dt]_{00}$ agreed to but a single significant figure, while those for $[d\chi/dt]_{00}$, $[di/dt]_{00}$, and $[d\Omega/dt]_{00}$ agreed to but two. It will be noticed that the increase of the number of points of division almost wholly removes the discrepancy.

Notwithstanding the entire disagreement of the test equations when but twelve points of division were employed, it is evident that this number would have been sufficient. The greatest error would have occurred with $[dL/dt]_{00}$, its amount being $0''.00000016$, showing that with this coefficient the sum of all terms of an order higher than the 12th amounts to but 1/4000000th of the remaining terms.

ACTION OF THE EARTH ON VENUS.

E	A	B cos ε	B sin ε	1000 × g	h
0°	1.53711333	+0.64403672	+0.34841048	0.03414346	1.00234463
30	1.52932317	+0.38337105	+0.61703357	0.10708849	1.0037255$\bar{4}$
60	1.51985488	+0.02331740	+0.71967967	0.14568124	1.0031407$\bar{0}$
90	1.51125435	−0.33964844	+0.62884478	0.11122751	1.00120600
120	1.50583081	−0.60827014	+0.36886855	0.03827088	0.9998313$\bar{1}$
150	1.50503256	−0.71057033	+0.00941083	0.00002491	1.0003553$\bar{1}$
180	1.50906482	−0.61913831	−0.35321171	0.03509098	1.0022640$\bar{2}$
210	1.51684286	−0.35847281	−0.62183471	0.10876150	1.0037093$\bar{5}$
240	1.52628659	+0.00158081	−0.72448100	0.14763156	1.0032674$\bar{7}$
270	1.53487476	+0.36454672	−0.63364618	0.11293247	1.0013352$\bar{7}$
300	1.54031063	+0.63316838	−0.37366983	0.03927365	0.99985008
330	1.54113325	+0.73546881	−0.01421211	0.00005681	1.00036791
Σ₁	9.13846106*	+0.07469486†	−0.01440384‡	0.44009177	6.01069819
Σ₂	9.13846095	+0.07469500	−0.01440382	0.44009169	6.01069936

E	l	G	G'	G''	θ		
					°	′	″
0°	0.53448743	1.00227181	0.53462398	0.00006372	46	55	3.02
30	0.5253163$\bar{7}$	1.00350237	0.52574251	0.00020298	46	22	33.07
60	0.5164329$\bar{2}$	1.00284204	0.51701255	0.00028098	45	53	55.40
90	0.50976708	1.00097979	0.51021108	0.00021779	45	33	45.30
120	0.5057182$\bar{4}$	0.99975382	0.50587139	0.00007567	45	20	44.51
150	0.5043959$\bar{9}$	1.00035525	0.50439608	0.00000005	45	14	29.83
180	0.5065195$\bar{4}$	1.00219338	0.50665928	0.00006911	45	19	11.90
210	0.5128522$\bar{5}$	1.00348844	0.51328431	0.00021116	45	39	53.44
240	0.5227378$\bar{6}$	1.00296095	0.52332564	0.00028127	46	15	21.69
270	0.5332582$\bar{3}$	1.00109414	0.53371072	0.00021137	46	54	18.41
300	0.54017928	0.99976461	0.54033745	0.00007270	47	19	22.49
330	0.54048407	1.00036779	0.54048430	0.00000011	47	18	38.53
Σ₁	3.12607525	6.00978661	3.12783029	0.00084345	277	3	39.01
Σ₂	3.12607397	6.00978778	3.12782900	0.00084346	277	3	38.58

* $6a^2 + 3a^2e^2 + 6[a'^2 - 2kaa'ee' \cos K] = 9.13846101.$

† $6[a'^2e' - kaa'e \cos K] = +0.07469471.$

‡ $-6k'aa' \cos \varphi' \cdot e \sin K' = -0.04440383.$

ACTION OF THE EARTH ON VENUS.

E	$\log K_0$	$\log L_0'$	$\log N_0$	$\log N$	$\log P$	$\log Q$
0°	0.26000336	0.60598257	0.54577154	9.8305330̄	0.4344894̄	0.3752914̄
30	0.25288171	0.59719127	0.53614961	9.8233230	0.4173018	0.3578664
60	0.24673133	0.58958546	0.52781896	9.8197364	0.4066135	0.3462012
90	0.24246622	0.58430371	0.52203035	9.8196999	0.4029640	0.3412105
120	0.23974361	0.58092896	0.51833025	9.8208350	0.4019120	0.3392393
150	0.23844519	0.57931865	0.51656434	9.8213596̄	0.4003696̄	0.3377696̄
180	0.23942219	0.58053038	0.51789317	9.8218870̄	0.4004544̄	0.3387987̄
210	0.24375790	0.58590391	0.52378440	9.8244982̄	0.4071947̄	0.3466789̄
240	0.25132635	0.59526901	0.53404476	9.8301980	0.4226554	0.3628370
270	0.25983876	0.60577958	0.54554439	9.8370023	0.4416487	0.3819851
300	0.26543283	0.61267386	0.55308969	9.8405753̄	0.4533906̄	0.3937357̄
330	0.26526804	0.61247091	0.55286781	9.8378789	0.4500302	0.3905869
Σ_1	1.50265967	3.56497024	3.19694837	8.9637646̄	2.5195152̄	2.1561032̄
Σ_2	1.50265782	3.56496803	3.19694594	8.9637618	2.5195089	2.1560973

E	$\log V$	J_1'	J_2	J_3	F_2
0°	0.3752596	0.9974825	+0.004202020	−0.047236424	−0.005832177
30	0.3577653	0.9964541	+0.010025286	−0.058127293	−0.010328762
60	0.3460609	0.9970986	+0.013405599	−0.053273840	−0.012046994
90	0.3411015	0.9987570	+0.012206904	−0.033976547	−0.010526476
120	0.3392013	0.9997616	+0.006516023	−0.005406098	−0.006174632
150	0.3377695	0.9991244	−0.001172845	+0.024782063	−0.000157532
180	0.3387641	0.9974877	−0.007608410	+0.048499056	+0.005912547
210	0.3465734	0.9964670	−0.010818504	+0.059389945	+0.010409132
240	0.3626968	0.9970681	−0.010873469	+0.054536468	+0.012127366
270	0.3818798	0.9987050	−0.008962465	+0.035239146	+0.010606846
300	0.3936994	0.9997505	−0.005884342	+0.006668715	+0.006255003
330	0.3905868	0.9991427	−0.001520954	−0.023519435	+0.000237902
Σ_1	2.1556821	5.9886490*	−0.000242579	+0.003787877	+0.000241113
Σ_2	2.1556763	5.9886502	−0.000242578	+0.003787879	+0.000241110

* $\Sigma_1(J_1' - G'') = 5.9878055̄.$

$\Sigma_2(J_1' - G'') = 5.9878067.$

ACTION OF THE EARTH ON VENUS.

E	$1000 \times F_3$	R_0	$1000 \times S_0$	W_0	$R^{(n)}$	$S^{(n)}$
$0°$	$+0.20122394$	0.42126107	-5.890097	-0.11153496	0.0000000	-0.008199111
30	$+0.06017487$	0.40675651	-4.150418	-0.13232119	$+0.2828447$	-0.005772122
60	-0.29485277	0.40441131	-0.984602	-0.11894044	$+0.4858527$	-0.001365876
90	-0.50823894	0.41101802	$+0.151227$	-0.07580674	$+0.5682286$	$+0.000209069$
120	-0.36563277	0.41646740	-1.349097	-0.01272800	$+0.4969244$	-0.001858754
150	-0.00856116	0.41403810	-2.948800	$+0.05391799$	$+0.2845157$	-0.004052670
180	$+0.20680805$	0.40710579	-1.730893	$+0.10632289$	0.0000000	-0.002376680
210	$+0.06559223$	0.40536093	$+2.554045$	$+0.13208012$	-0.2785530	$+0.003510140$
240	-0.29105383	0.41526071	$+7.029019$	$+0.12494379$	-0.4954847	$+0.009684416$
270	-0.50707628	0.43289840	$+7.732079$	$+0.08349760$	-0.5984781	$+0.010689527$
300	-0.36741788	0.44451245	$+3.199250$	$+0.01546615$	-0.5340294	$+0.004438118$
330	-0.01281582	0.43893384	-3.068000	-0.05784756	-0.3052198	-0.004266768
Σ_1	-0.91092526	2.50901873	$+0.273580$	$+0.00352943$	-0.0467370	$+0.000322113$
Σ_2	-0.91092510	2.50900580	$+0.270133$	$+0.00352022$	-0.0466619	$+0.000317176$

E	$[R_0 \sin v + (\cos v + \cos E)S_0]$	$\left[-R_0 \cos v + \left(\frac{r}{a}\sec^2\varphi + 1\right)\sin v S_0\right]$	$W_0 \cos u$	$W_0 \sin u$	$-2\frac{r}{a}R_0$
$0°$	-0.01178019	-0.42126107	-0.065352039	-0.09038273	-0.83675673
30	$+0.19740436$	-0.35572418	-0.013079792	-0.13167315	-0.80869188
60	$+0.35044516$	-0.20183130	$+0.049268295$	-0.10825646	-0.80605518
90	$+0.41100745$	$+0.00311509$	$+0.061732775$	-0.04399687	-0.82203604
120	$+0.36078900$	$+0.20803111$	$+0.012668751$	-0.00122668	-0.83578481
150	$+0.21090709$	$+0.35633148$	-0.049130989	-0.02221027	-0.83298365
180	$+0.00346179$	$+0.40710579$	-0.062298971	-0.08615914	-0.81978340
210	-0.20590981	$+0.34919573$	-0.013955080	-0.13134085	-0.81552660
240	-0.36545661	$+0.19760050$	$+0.050403535$	-0.11432600	-0.83336328
270	-0.43294230	-0.01250179	$+0.067326200$	-0.04938656	-0.86579680
300	-0.38308894	-0.22551772	$+0.015375410$	-0.00167292	-0.88598306
330	-0.22607885	-0.37629530	-0.052873488	-0.02346769	-0.87266531
Σ_1	-0.04562979	-0.03587269	$+0.000064081$	-0.40202393	-5.01772646
Σ_2	-0.04561206	-0.03587897	$+0.000019626$	-0.40207539	-5.01770028

$$\sin\varphi \cdot \tfrac{1}{2}A_1^{(s)} + \cos\varphi \cdot B_0^{(c)} = +0.00000000814.$$

DIFFERENTIAL COEFFICIENTS.

				log coeff.
$[de/dt]_{00}$	$= -$	16017.410	m'	$n\ 4.2045923$
$[d\chi/dt]_{00}$	$= -$	1840673.3	m'	$n\ 6.2649767$
$[di/dt]_{00}$	$= +$	14.695	m'	$p\ 1.1671802$
$[d\Omega/dt]_{00}$	$= -$	2385136.3	m'	$n\ 6.3775132$
$[d\pi/dt]_{00}$	$= -$	1844854.1	m'	$n\ 6.2659621$
$[dL/dt]_{00}$	$= -$	1765973.3	m'	$n\ 6.2469841$

FINAL VALUES CORRESPONDING TO THE ABOVE VALUE OF m'.

$$[de/dt]_{00} = -0.04898290$$
$$[d\chi/dt]_{00} = -5.6289701$$
$$[di/dt]_{00} = +0.000044940$$
$$[d\Omega/dt]_{00} = -7.293993$$
$$[d\pi/dt]_{00} = -5.6417558$$
$$[dL/dt]_{00} = -5.4005288$$

COMPARISON WITH OTHER RESULTS.

	Leverrier.	Newcomb.	Method of Gauss.
$[de/dt]_{00}$	-0.04875	-0.04896	-0.048982
$e[d\pi/dt]_{00}$	-0.03873	-0.03852	-0.038607
$[di/dt]_{00}$	$+0.00006$	$+0.00004$	$+0.0000449$
$\sin i\ [d\Omega/dt]_{00}$	-0.43154	-0.43169	-0.431698
$[dL/dt]_{00}$	-5.397		-5.4005

NOTES.

The close agreement of the sums of the functions toward the beginning of the computation is here caused by the smallness of the term a'^2e'; the ratio of the major axes is, however, so large that the expansion of the perturbing function is not very rapidly convergent. The greatest error arising from a division of the orbit into but six parts would here occur with the coefficient $[d\Omega/dt]_{00}$, its amount being $0''.0004$, which is 1/16000th of the whole.

ACTION OF MARS ON VENUS.

E	A	$B \cos \epsilon$	$B \sin \epsilon$	g	h
0°	2.6510232	−0.78427108	+0.4410282	0.003928189	2.30246275
30	2.6356206	−0.87195528	−0.1200582	0.000291101	2.30200670
60	2.6766222	−0.66613155	−0.6497891	0.008527164	2.30404590
90	2.7630522	−0.22195070	−1.0062238	0.020447915	2.30677230
120	2.8717556	+0.34156979	−1.0938564	0.024164644	2.30762720
150	2.9736020	+0.87343460	−0.8892048	0.015968474	2.30563865
180	3.0412895	+1.23113141	−0.4471057	0.004037199	2.30274975
210	3.0566810	+1.31881567	+0.1139805	0.000262374	2.30215960
240	3.0156549	+1.11299237	+0.6437115	0.008368400	2.30456350
270	2.9292114	+0.66881112	+1.0001462	0.020201647	2.30723845
300	2.8205208	+0.10529076	+1.0877783	0.023896846	2.30739440
330	2.7187000	−0.42657423	+0.8831271	0.015750931	2.30502860
Σ_1	17.0768662*	+1.34058170†	−0.0182332‡	0.072922442	13.82884350
Σ_2	17.0768672	+1.34058118	−0.0182329	0.072922442	13.82884430

E	l	G	G'	G''	θ		
					°	′	″
0	0.32836475	2.30159781	0.33433455	0.00510484	22	33	26.00
30	0.31341820	2.30194311	0.31388465	0.00040288	21	40	58.40
60	0.35238060	2.30214618	0.36444375	0.01016345	23	44	4.25
90	0.43608420	2.30201186	0.46014840	0.01930383	27	1	51.11
120	0.54393270	2.30165423	0.56837725	0.01847156	30	11	38.49
150	0.64776765	2.30144287	0.66243750	0.01047414	32	38	59.04
180	0.71834405	2.30164190	0.72188175	0.00242983	34	6	10.10
210	0.73432570	2.30208690	0.73455360	0.00015516	34	23	46.69
240	0.69089570	2.30230785	0.69835610	0.00520478	33	30	59.53
270	0.60177725	2.30207736	0.62106785	0.01412953	31	34	45.83
300	0.49293070	2.30165418	0.51868770	0.02001680	28	47	46.93
330	0.39347570	2.30144156	0.41360955	0.01654686	25	31	1.93
Σ_1	3.12684850	13.81100215	3.20608110	0.06139126	172	54	5.30
Σ_2	3.12684870	13.81100366	3.20570155	0.06101240	172	51	23.00

* $6a^2 + 3a^2e^2 + 6[a'^2 - 2kaa'ee' \cos K] = 17.0768669$.

† $6[a'^2e' - kaa'e \cos K] = +1.34058156$.

‡ $-6k'aa' \cos \varphi' \cdot e \sin K' = -0.0182328$.

ACTION OF MARS ON VENUS.

E	$\log K_0$	$\log L_0'$	$\log N_0$	$\log N$	$\log P$	$\log Q$
0°	0.05236235	0.34221622	0.25376720	9.0799241	8.6961571	8.9706997
30	0.04824440	0.36381569	0.24772073	9.0778391	8.6903136	8.9633892
60	0.05820469	0.34986592	0.26232763	9.0871721	8.7089460	8.9854537
90	0.07644597	0.37365893	0.28892064	9.1058581	8.7480486	9.0290445
120	0.09671054	0.39993150	0.31822634	9.129423$\overline{7}$	8.7983322	9.082138$\overline{5}$
150	0.11444649	0.42279102	0.34867356	9.1516340	9.8464806	9.1313354
180	0.12581914	0.43738414	0.35989311	9.1660121	8.8784042	9.1634092
210	0.12819869	0.44043118	0.36327720	9.168118$\overline{2}$	8.884247$\overline{6}$	9.1692445
240	0.12114959	0.43139835	0.35324262	9.157413$\overline{9}$	8.8625240	9.1475124
270	0.10649224	0.41255431	0.33228403	9.1373395	8.8203390	9.1048461
300	0.08740987	0.38789402	0.30480686	9.113745$\overline{3}$	8.7700379	9.052751$\overline{5}$
330	0.06771795	0.36229172	0.27622192	9.0929018	8.7249711	9.0040125
Σ_1	0.54165618	2.34869015	1.85226376	4.733691$\overline{0}$	2.7144014	4.4019649
Σ_2	0.54154574	2.34854285	1.85209808	4.733690$\overline{7}$	2.714400$\overline{5}$	4.4018722

E	$\log V$	J_1'	J_2	J_3	F_2
0°	8.9695199	2.3059508	+0.040158678	−0.039429682	−0.09502735
30	8.9632959	2.2999322	−0.012465787	−0.068607310	+0.02586866
60	8.9831123	2.3089841	−0.060962061	−0.080069400	+0.14000859
90	9.0246325	2.3187331	−0.093160848	−0.070744754	+0.21680880
120	9.0779450	2.3192082	−0.101285499	−0.043131936	+0.23569076
150	9.1289682	2.3119133	−0.083189540	−0.004629729	+0.19159498
180	9.1628614	2.3032758	−0.042922865	+0.034445271	+0.09633687
210	9.1692095	2.2997005	+0.009559512	+0.063622976	−0.02455913
240	9.1463381	2.3040271	+0.060251438	+0.075085208	−0.13869908
270	9.1016472	2.3135287	+0.094787774	+0.065760616	−0.21549926
300	9.0481938	2.3207258	+0.103054621	+0.038147729	−0.23438118
330	9.0002162	2.3179852	+0.082763214	−0.000354612	−0.19028541
Σ_1	4.3879705	13.8621718*	−0.001705675	−0.014952810	+0.00392861
Σ_2	4.3879695	13.8617930	−0.001705688	−0.014952813	+0.00392864

* $\Sigma_1(J_1' - G'') = 13.800780\overline{6}.$
$\Sigma_2(J_1' - G'') = 13.8007806.$

ACTION OF MARS ON VENUS.

E	$1000 \times F_3$	R_0	$1000 \times S_0$	$1000 \times W_0$	$R^{(n)}$	$1000 \times S^{(n)}$
0°	+ 2.8021516	0.06350844	− 0.9770071	− 3.5365233	0.00000000	− 1.3600097
30	− 0.4518739	0.06287361	+ 0.1223535	− 6.3268753	+ 0.04372018	+ 0.1701610
60	− 0.0998524	0.06461988	+ 1.2993923	− 7.7066718	+ 0.07763320	+ 1.8025651
90	+ 3.5312016	0.06860530	+ 2.2776450	− 7.2896438	+ 0.09484620	+ 3.1488232
120	+ 6.8298095	0.07412522	+ 2.6943730	− 4.7318401	+ 0.08844541	+ 3.7122436
150	+ 6.5062493	0.07970667	+ 2.2590820	− 0.1661617	+ 0.05477227	+ 3.1047600
180	+ 2.8799133	0.08340064	+ 1.0358414	+ 5.2294311	0.00000000	+ 1.4223079
210	− 0.4389788	0.08380114	− 0.4699266	+ 9.3597858	− 0.05758587	− 0.6458416
240	− 0.1552801	0.08095204	− 1.6671420	+ 10.5057004	− 0.09659109	− 2.2969487
270	+ 3.4223024	0.07610223	− 2.2705920	+ 8.5365166	− 0.10521063	− 3.1390719
300	+ 6.6966141	0.07080065	− 2.2876620	+ 4.6568417	− 0.08505867	− 3.1735299
330	+ 6.3844559	0.06631737	− 1.8208410	+ 0.3034394	− 0.04611486	− 2.5323023
Σ_1	+ 18.9533560	0.43740687	− 0.0977956	+ 4.4169380	− 0.01557115	+ 0.1066283
Σ_2	+ 18.9533565	0.43740632	− 0.0977209	+ 4.4170612	− 0.01557271	+ 0.1065284

E	$[R_0 \sin v + (\cos v + \cos E)S_0]$	$\left[-R_0 \cos v + \left(\dfrac{r}{a}\sec^2\varphi + 1\right)\sin v S_0\right]$	$1000 \times W_0 \cos u$	$1000 \times W_0 \sin u$	$-2\dfrac{r}{a}R_0$
0°	− 0.001954014	− 0.06350844	− 2.072195	− 2.865834	− 0.12614768
30	+ 0.031835191	− 0.05421922	− 0.625404	− 6.295890	− 0.12500201
60	+ 0.057445987	− 0.02972268	+ 3.192309	− 7.014411	− 0.12879747
90	+ 0.068588125	+ 0.00502476	+ 5.936279	− 4.230780	− 0.13721060
120	+ 0.061265775	+ 0.04210058	+ 4.709813	− 0.456037	− 0.14875771
150	+ 0.035700934	+ 0.07141598	+ 0.151409	+ 0.068446	− 0.16035811
180	− 0.002071682	+ 0.08340064	− 3.064140	− 4.237688	− 0.16794274
210	− 0.040838000	+ 0.07318497	− 0.988919	− 9.307396	− 0.16859554
240	− 0.068190156	+ 0.04377273	+ 4.238101	− 9.612920	− 0.16245809
270	− 0.076084930	+ 0.00506196	+ 6.883205	− 5.049116	− 0.15220446
300	− 0.063800118	− 0.03106656	+ 4.629519	− 0.503716	− 0.14111682
330	− 0.036506252	− 0.05549212	+ 0.277348	+ 0.123100	− 0.13184872
Σ_1	− 0.017304208	+ 0.04497627	+ 11.633407	− 24.690606	− 0.87522051
Σ_2	− 0.017304932	+ 0.04497633	+ 11.633918	− 24.691636	− 0.87521944

$$\sin\varphi \cdot \tfrac{1}{2}A_1^{(s)} + \cos\varphi \cdot B_0^{(c)} = + 0.0000000018.$$

DIFFERENTIAL COEFFICIENTS.

			log coeff.
$[de/dt]_{00}$	$= -$	$6075.5972\ m'$	$n\ 3.7835890$
$[d\chi/dt]_{00}$	$= +2307588.8$	m'	$p\ 6.3631584$
$[di/dt]_{00}$	$= +$	$4084.7434\ m'$	$p\ 3.6111648$
$[d\Omega/dt]_{00}$	$= -$	$146478.61\ m'$	$n\ 5.1657742$
$[d\pi/dt]_{00}$	$= +2307332.0$	m'	$p\ 6.3631101$
$[dL/dt]_{00}$	$= -$	$307497.75\ m'$	$n\ 5.4878419$

FINAL VALUES CORRESPONDING TO THE ABOVE VALUE OF m'.

$$[de/dt]_{00} = -0.\overset{''}{0}019639882$$
$$[d\chi/dt]_{00} = +0.74594759$$
$$[di/dt]_{00} = +0.0013204280$$
$$[d\Omega/dt]_{00} = -0.047350446$$
$$[d\pi/dt]_{00} = +0.74586465$$
$$[dL/dt]_{00} = -0.099401232$$

COMPARISON WITH OTHER RESULTS.

	Leverrier.	Newcomb.	Method of Gauss.
$[de/dt]_{00}$	-0.00195	-0.00196	-0.001964
$e[d\pi/dt]_{00}$	$+0.00510$	$+0.00510$	$+0.005104$
$[di/dt]_{00}$	$+0.00131$	$+0.00132$	$+0.001320$
$\sin i\ [d\Omega/dt]_{00}$	-0.00280	-0.00281	-0.002802
$[dL/dt]_{00}$	-0.099		-0.099401

NOTES.

The close agreement of the final sums shows that, notwithstanding the high eccentricity of the orbit of Mars, the expansion of the perturbing function is quite rapidly convergent for this case. The greatest error arising from a division into but six parts would here occur with the coefficient $[d\Omega/dt]_{00}$, and would amount to 1/50000th of the whole value of this coefficient.

ACTION OF JUPITER ON VENUS.

E	A	$B \cos \epsilon$	$B \sin \epsilon$	g	h
0°	27.41848845	−0.4215334	+3.3076466	0.68960365	27.00777$\overline{9}$
30	27.28099847	−1.8560261	+1.9912702	0.24993195	27.00726$\overline{6}$
60	27.22724715	−2.4400062	+0.1352278	0.00115264	27.006632
90	27.27164617	−2.0169963	−1.7631551	0.19594869	27.00651$\overline{6}$
120	27.40230326	−0.7003432	−3.1952096	0.64351702	27.007039
150	27.58420446	+1.1571588	−3.7772161	0.89930104	27.00770$\overline{1}$
180	27.76860046	+3.0577930	−3.3532271	0.70874081	27.00785$\overline{1}$
210	27.90607834	+4.4922856	−2.0368516	0.26150512	27.007330
240	27.95980514	+5.0762663	−0.1808092	0.00206064	27.006665
270	27.91539376	+4.6532564	+1.7175743	0.18594838	27.006520
300	27.78474891	+3.3366031	+3.1496275	0.62528757	27.007025
330	27.60287233	+1.4791014	+3.7316353	0.87772776	27.007658
Σ_1	165.56119337*	+7.9087796†	−0.1367440‡	2.67036233	162.042990
Σ_2	165.56119353	+7.9087799	−0.1367430	2.67036294	162.04299$\overline{0}$

E	l	G	G'	G''	θ ° ′ ″
0	0.34767$\overline{8}$	27.0068207	0.4107944	0.06215867	7 35 44.57
30	0.21070$\overline{1}$	27.0069201	0.2483150	0.03726867	5 53 53.53
60	0.157583	27.0066304	0.1578551	0.00027037	4 23 18.41
90	0.20209$\overline{9}$	27.0062448	0.2334496	0.03108029	5 40 35.79
120	0.332232	27.0061457	0.393656$\overline{8}$	0.06053126	7 26 34.62
150	0.51347$\overline{2}$	27.0064436	0.572857$\overline{6}$	0.05812876	8 46 58.16
180	0.69771$\overline{8}$	27.0068530	0.7344471	0.03573164	9 42 56.80
210	0.835716	27.0069600	0.847511$\overline{4}$	0.01142507	10 16 14.68
240	0.890108	27.0066621	0.8901968	0.00008571	10 27 38.97
270	0.845842	27.0062568	0.854165$\overline{9}$	0.00806094	10 17 29.01
300	0.714692	27.0061444	0.7465850	0.03101258	9 45 50.06
330	0.532182	27.0064305	0.5886244	0.05521465	8 52 22.88
Σ_1	3.140010	162.0392563	3.3335352	0.18979023	49 22 3.43
Σ_2	3.14001$\overline{1}$	162.0392558	3.344923$\overline{8}$	0.20117838	49 47 34.05

* $6a^2 + 3a^2e^2 + 6[a'^2 − 2kaa'ee' \cos K] = 165.5611934.$

† $6[a'^2e' − kaa'e \cos K] = +7.9087800.$

‡ $− 6k'aa' \cos \varphi' \cdot e \sin K' = − 0.1367433.$

ACTION OF JUPITER ON VENUS.

E	$\log K_0$	$\log L_0'$	$\log N_0$	$\log N$	$\log P$	$\log Q$
0°	0.00574768	0.28065745	0.18470221	7.429089$\bar{1}$	4.8448027	6.181319$\bar{4}$
30	0.00346017	0.27761215	0.18127767	7.428200$\bar{0}$	4.8416643	6.177403$\bar{8}$
60	0.00191340	0.27555166	0.17896019	7.429737$\bar{8}$	4.842340$\bar{1}$	6.177223$\bar{3}$
90	0.00320448	0.27727161	0.18089469	7.433272$\bar{4}$	4.846616$\bar{6}$	6.1822034
120	0.00551791	0.28035168	0.18435840	7.437846$\bar{0}$	4.853327$\bar{9}$	6.189769$\bar{5}$
150	0.00769513	0.28324824	0.18761506	7.4422392	4.8606851	6.197453$\bar{2}$
180	0.00942826	0.28555246	0.19020526	7.4452930	4.8667491	6.2034501
210	0.01054435	0.28703562	0.19187230	7.446201$\bar{1}$	4.869917$\bar{9}$	6.206414$\bar{0}$
240	0.01094130	0.28756298	0.19246500	7.4447133	4.8693318	6.205706$\bar{1}$
270	0.01058711	0.28709242	0.19193614	7.4412095	4.8651139	6.2015516
300	0.00952253	0.28567775	0.19034610	7.4366174	4.858373$\bar{2}$	6.1950025
330	0.00785489	0.28346069	0.18785389	7.4321742	4.8509265	6.1876739
Σ_1	0.04307108	1.69535398	1.12103716	4.623296$\bar{5}$	9.134924$\bar{7}$	7.1524707
Σ_2	0.04334613	1.69572073	1.12144975	4.623296$\bar{3}$	9.1349242	7.152699$\bar{8}$

E	$\log V$	J_1'	J_2	J_3	F_2
0°	6.1800737	27.032601604	+0.14514043	−0.9796507	−4.3121624
30	6.1766560	27.002428557	+0.10192507	−1.0502482	−2.5960090
60	6.1772178	26.980449138	+0.02691302	−0.8390983	−0.1762960
90	6.1815797	27.031356993	−0.07055224	−0.4027785	+2.2986164
120	6.188556$\bar{3}$	27.065863046	−0.15999110	+0.1417995	+4.1655781
150	6.196289$\bar{1}$	27.048572013	−0.20267009	+0.6487166	+4.9243381
180	6.2027348	27.006174576	−0.17674090	+0.9821448	+4.3715859
210	6.206185$\bar{3}$	26.976509919	−0.09316228	+1.0527424	+2.6554331
240	6.2057043	26.979786076	+0.01124440	+0.8415929	+0.2357201
270	6.2013902	27.007935305	+0.09774464	+0.4052732	−2.2391928
300	6.1943816	27.036474318	+0.14682513	−0.1393049	−4.1061538
330	6.1865682	27.045934111	+0.16010585	−0.6462222	−4.8649146
Σ_1	7.148668$\bar{5}$	162.101348758*	−0.00660902	+0.0074833	+0.1782719
Σ_2	7.1486684	162.112736898	−0.00660905	+0.0074833	+0.1782712

* $\Sigma_1(J_1' - G'') = 161.911558528.$
$\Sigma_2(J_1' - G'') = 161.911558518.$

ACTION OF JUPITER ON VENUS.

E	F_3	$1000 \times R_0$	$100000 \times S_0$	$1000 \times W_0$	$1000 \times R^{(n)}$	$100000 \times S^{(n)}$
0°	+0.06417143	1.3439836	−0.81929943	−0.14785238	0.0000000	−1.1404782
30	−0.01379964	1.3378723	−0.27202988	−0.15783804	+0.9303113	−0.3783208
60	−0.00425156	1.3439282	+0.28211779	−0.12622122	+1.6145721	+0.3913643
90	+0.08385809	1.3588647	+0.54292475	−0.06059607	+1.8786178	+0.7505883
120	+0.16287446	1.3765538	+0.50194299	+0.02305118	+1.6424891	+0.6915652
150	+0.15397825	1.3917093	+0.38822446	+0.10305711	+0.9563448	+0.5335546
180	+0.06595223	1.4018961	+0.39768404	+0.15712801	0.0000000	+0.5460578
210	−0.01357095	1.4066592	+0.47042587	+0.16914113	−0.9666180	+0.6465278
240	−0.00563629	1.4053074	+0.35504098	+0.13510528	−1.6767973	+0.4891671
270	+0.08123098	1.3964225	−0.08725072	+0.06503287	−1.9305409	−0.1206233
300	+0.15970886	1.3800824	−0.66641223	−0.02064190	−1.6580071	−0.9244718
330	+0.15112247	1.3602249	−0.99122256	−0.09822805	−0.9458544	−1.3785254
Σ_1	+0.44281913	8.2517515	+0.05107414	+0.02056897	−0.0777432	+0.0532044
Σ_2	+0.44281920	8.2517529	+0.05107192	+0.02056895	−0.0777394	+0.0532012

E	$1000 \times [R_0 \sin v + (\cos v + \cos E)S_0]$	$1000 \times \left[-R_0 \cos v + \left(\dfrac{r}{a}\sec^2\varphi + 1\right)\sin v \cdot S_0 \right]$	$1000 \times W_0 \cos u$	$1000 \times W_0 \sin u$	$-2\dfrac{r}{a}R_0$
0°	−0.0163860	−1.3439836	−0.086632804	−0.11981268	−0.0026695736
30	+0.6682014	−1.1590573	−0.015602101	−0.15706504	−0.0026598877
60	+1.1706512	−0.6601482	+0.052284193	−0.11488325	−0.0026786592
90	+1.3587959	+0.0201574	+0.049346057	−0.03516889	−0.0027177294
120	+1.1829926	+0.7039968	−0.022943878	+0.00222159	−0.0027625280
150	+0.6850081	+1.2114930	−0.093907380	−0.04245200	−0.0027999142
180	−0.0079537	+1.4018961	−0.092067787	−0.12732924	−0.0028229786
210	−0.7073256	+1.2159042	−0.017870798	−0.16819438	−0.0028299915
240	−1.2164221	+0.7037025	+0.054502775	−0.12362395	−0.0028202318
270	−1.3963839	+0.0113009	+0.052437627	−0.03846518	−0.0027928449
300	−1.2058917	−0.6713715	−0.020520787	+0.00223277	−0.0027507212
330	−0.7013026	−1.1657067	−0.089781837	−0.03984932	−0.0027043278
Σ_1	−0.0930097	+0.1340921	−0.115378288	−0.48119476	−0.0165046924
Σ_2	−0.0930067	+0.1340915	−0.115378432	−0.48119481	−0.0165046855

$$\sin\varphi \cdot \tfrac{1}{2}A_1^{(s)} + \cos\varphi \cdot B_0^{(c)} = +0.0000000000028.$$

DIFFERENTIAL COEFFICIENTS.

			log coeff.
$[de/dt]_{00}$	$= -$	$32.\overset{''}{6}54970 \; m'$	$n \; 1.5139493$
$[d\chi/dt]_{00}$	$= +6879.8159$	m'	$p \; 3.8375768$
$[di/dt]_{00}$	$= -$	$40.510972 \; m'$	$n \; 1.6075727$
$[d\Omega/dt]_{00}$	$= -2854.6599$	m'	$n \; 3.4555544$
$[d\pi/dt]_{00}$	$= +6874.8117$	m'	$p \; 3.8372608$
$[dL/dt]_{00}$	$= -5799.7390$	m'	$n \; 3.7634084$

FINAL VALUES CORRESPONDING TO THE ABOVE VALUE OF m'.

$$[de/dt]_{00} = -0.\overset{''}{0}31162921$$
$$[d\chi/dt]_{00} = +6.5654682$$
$$[di/dt]_{00} = -0.038659982$$
$$[d\Omega/dt]_{00} = -2.7242270$$
$$[d\pi/dt]_{00} = +6.5606924$$
$$[dL/dt]_{00} = -5.5347410$$

COMPARISON WITH OTHER RESULTS.

	Leverrier.	Newcomb.	Method of Gauss.
$[de/dt]_{00}$	$-0.\overset{''}{0}3117$	$-0.\overset{''}{0}3117$	$-0.\overset{''}{0}311629$
$e[d\pi/dt]_{00}$	$+0.04482$	$+0.04491$	$+0.0448955$
$[di/dt]_{00}$	-0.03865	-0.03865	-0.0386600
$\sin i \; [d\Omega/dt]_{00}$	-0.16114	-0.16122	-0.1612345
$[dL/dt]_{00}$	-5.535		-5.5347410

NOTES.

The term $a^2 e'$ is here so large that the sums of the functions B, ϵ, G', G'', and θ, as well as those of the functions immediately dependent upon these quantities are in great disagreement; but, as the expansion of the perturbing function is here rapidly convergent, the final sums agree almost exactly. The greatest effect of all terms from the 6th to the 11th orders is here produced with the coefficient $[di/dt]_{00}$ and amounts to $0''.00000002$.

ACTION OF SATURN ON VENUS.

E	A	B cos ε	B sin ε	g	h
0°	92.09886822	+10.392665	+4.3401600	5.386574	90.704833
30	91.77434432	+ 7.489741	+6.4166597	11.773870	90.705340̄
60	91.37861335	+ 3.936959	+6.7658078	13.090030	90.705178
90	91.01772022	+ 0.686282	+5.2940524	8.014522	90.704527
120	90.78837022	− 1.391274	+2.3957451	1.641279	90.704015
150	90.75201312	− 1.739027	−1.1525126	0.379833	90.704174
180	90.91838168	− 0.263800	−4.3999687	5.536055	90.704848
210	91.24289345	+ 2.639122	−6.4764687	11.994380	90.705369
240	91.63859982	+ 6.191904	−6.8256188	13.322485	90.705222
270	91.99948069	+ 9.442580	−5.3538604	8.196630	90.704548
300	92.22884291	+11.520135	−2.4555545	1.724250	90.704026
330	92.26522467	+11.867891	+1.0927030	0.341433	90.704165
Σ_1	549.05167620*	+30.386589†	−0.1794291‡	40.700673	544.228122
Σ_2	549.05167647	+30.386589	−0.1794266	40.700668	544.228123

E	l	G	G′	G″	θ		
					°	′	″
0	+1.108078	90.7041702	1.1599384	0.0511977	6	38	1.200
30	+0.783048̄	90.7038960	0.9248448	0.1403539	6	12	59.177
60	+0.387477	90.7035801	0.6213419̄	0.2322660	5	33	35.601
90	+0.027235	90.7035526	0.3116929	0.2834826	4	38	20.618
120	−0.201603	90.7038159	0.0673334	0.2687363	3	29	4.472
150	−0.238118	90.7041279	0.0164527̄	0.2545245	3	7	43.816
180	−0.072423	90.7041756	0.2137666	0.2855179	4	14	53.293
210	+0.251567	90.7039071	0.5115373	0.2585083	5	16	44.963
240	+0.647420	90.7035910	0.8267173	0.1776657	6	2	4.051
270	+1.008976	90.7035405	1.0926850̄	0.0827021̄	6	32	0.533
300	+1.238859	90.7038135	1.2542286	0.0151565	6	47	36.301
330	+1.275102	90.7041229	1.2780897	0.0029452	6	49	30.415
Σ_1	+3.107808	544.2231463	4.1433262̄	1.0305401	32	45	14.918
Σ_2	+3.107809̄	544.2231470	4.1353023	1.0225166̄	32	37	19.522

* $6a^2 + 3a^2e^2 + 6[a'^2 − 2kaa'ee' \cos K] = 549.05167622$.

† $6[a'^2e' − kaa'e \cos K] = + 30.386587$.

‡ $− 6k'aa' \cos \varphi' \cdot e \sin K' = − 0.1794290$.

Action of Saturn on Venus.

E	$\log K_0$	$\log L_0{}'$	$\log N_0$	$\log N$	$\log P$	$\log Q$
0°	0.00437971	0.27883659	0.18265468	6.6396204	3.0027123	4.864402$\overline{8}$
30	0.00384466	0.27812417	0.18185352	6.6392492	3.0007786	4.8628053
60	0.00307377	0.27709752	0.18069890	6.640007$\overline{6}$	2.999634$\overline{9}$	4.861971$\overline{4}$
90	0.00213852	0.27585161	0.17929758	6.6416827	2.9995754	4.862000$\overline{9}$
120	0.00120578	0.27460866	0.17789947	6.6438207	3.0006087	4.862809$\overline{9}$
150	0.00097200	0.27429705	0.17754895	6.6458518	3.002460$\overline{9}$	4.864556$\overline{8}$
180	0.00179285	0.27539102	0.17877951	6.647241$\overline{5}$	3.004648$\overline{0}$	4.8670287
210	0.00277061	0.27669370	0.18024473	6.6476233	3.0065929	4.869006$\overline{0}$
240	0.00362228	0.27782805	0.18152050	6.6468912	3.007770$\overline{6}$	4.869937$\overline{4}$
270	0.00424801	0.27866126	0.18245753	6.6452313	3.007852$\overline{5}$	4.869668$\overline{8}$
300	0.00459387	0.27912172	0.18297533	6.6430830	3.0068083	4.8683601
330	0.00463699	0.27917912	0.18303987	6.6410259	3.0049227	4.866424$\overline{7}$
Σ_1	0.01866826	1.66288356	1.08452839	9.8606643	8.022182$\overline{7}$	9.1945101
Σ_2	0.01861079	1.66280691	1.08444218	9.8606642	8.0221829	9.194462$\overline{3}$

E	$\log V$	$J_1{}'$	J_2	J_3	F_2
0°	4.8640969	90.64262161	$+0.26581387$	-3.1529304	-22.089684
30	4.8619670	90.77821079	$+0.41797337$	-2.4126056	-32.658241
60	4.8605842	90.92403274	$+0.41481456$	-1.0167115	-34.435270
90	4.860307$\overline{6}$	90.98237184	$+0.27423234$	$+0.6607234$	-26.944615
120	4.8612040	90.92121232	$+0.07649557$	$+2.1702330$	-12.193388
150	4.8630356	90.85355551	-0.10022240	$+3.1073439$	$+\ 5.865829$
180	4.865323$\overline{1}$	90.87694181	-0.22507782	$+3.2209585$	$+22.394088$
210	4.867462$\overline{1}$	90.89557561	-0.30595380	$+2.4806343$	$+32.962649$
240	4.8688763	90.86858755	-0.34689500	$+1.0847396$	$+34.739680$
270	4.869174$\overline{8}$	90.78219477	-0.32168268	-0.5926962	$+27.249019$
300	4.868269$\overline{6}$	90.66900008	-0.19522740	-2.1022048	$+12.497792$
330	4.866407$\overline{1}$	90.60246403	$+0.02557719$	-3.0393154	$-\ 5.561423$
Σ_1	9.1883540	544.90239611*	-0.01007622	$+0.2040844$	$+\ 0.913218$
Σ_2	9.188354$\overline{1}$	544.89437255	-0.01007598	$+0.2040844$	$+\ 0.913218$

* $\Sigma_1(J_1{}' - G'') = 543.87185601.$
$\Sigma_2(J_1{}' - G'') = 543.87185600.$

ACTION OF SATURN ON VENUS.

E	F_3	$1000 \times R_0$	$100000 \times S_0$	$1000 \times W_0$	$1000 \times R^{(n)}$	$100000 \times S^{(n)}$
$0°$	-0.15178108	0.21824849	-0.02789044	-0.023072725	0.00000000	-0.03882394
30	-0.76855930	0.21811010	-0.02299898	-0.017634117	$+0.15166640$	-0.03198543
60	-1.16657105	0.21852784	-0.04315220	-0.007491896	$+0.26253554$	-0.05986233
90	-0.94659087	0.21909707	-0.07037836	$+0.004695343$	$+0.30289965$	-0.09729742
120	-0.32871848	0.21963056	-0.06653491	$+0.015732681$	$+0.26206084$	-0.09167023
150	$+0.06775179$	0.22023472	-0.01412236	$+0.022675427$	$+0.15133936$	-0.01940900
180	-0.15599309	0.22103492	$+0.06128516$	$+0.023605169$	0.00000000	$+0.08415031$
210	-0.77884429	0.22186276	$+0.10918347$	$+0.018202966$	-0.15245807	$+0.15005585$
240	-1.18017310	0.22225231	$+0.09717617$	$+0.007900358$	-0.26518905	$+0.13388702$
270	-0.95986467	0.22180036	$+0.03944793$	-0.004483114	-0.30663694	$+0.05453639$
300	-0.33810816	0.22057741	-0.01719598	-0.015556236	-0.26499777	-0.02385490
330	$+0.06476239$	0.21916657	-0.03744378	-0.022338577	-0.15240100	-0.05207429
Σ_1	-3.32134496	1.32027153	$+0.00368780$	$+0.001117351$	-0.00559044	$+0.00382593$
Σ_2	-3.32134495	1.32027158	$+0.00368792$	$+0.001117928$	-0.00559070	$+0.00382610$

E	$1000 \times [R_0 \sin v + (\cos v + \cos E)S_0]$	$1000 \times \left[-R_0 \cos v + \left(\frac{r}{a}\sec^2\varphi + 1\right)\sin vS_0 \right]$	$1000 \times W_0 \cos u$	$1000 \times W_0 \sin u$	$1000 \times -2\frac{r}{a}R_0$
$0°$	-0.00055781	-0.21824849	-0.013519262	-0.018697064	-0.43351010
30	$+0.10930468$	-0.18874417	-0.001743111	-0.017547753	-0.43363510
60	$+0.18946670$	-0.10888722	$+0.003103342$	-0.006818927	-0.43556030
90	$+0.21909678$	$+0.00009174$	-0.003823625	$+0.002725094$	-0.43819414
120	$+0.19022138$	$+0.10978821$	-0.015659442	$+0.001516257$	-0.44076418
150	$+0.10971092$	$+0.19096268$	-0.020662231	-0.009340619	-0.44307990
180	-0.00122570	$+0.22103492$	-0.013831594	-0.019129016	-0.44509495
210	-0.11216824	$+0.19142744$	-0.001923256	-0.018101079	-0.44635520
240	-0.19279204	$+0.11058265$	$+0.003187081$	-0.007228982	-0.44602556
270	-0.22179790	$+0.00072885$	-0.003614847	$+0.002651640$	-0.44360072
300	-0.19184805	-0.10885437	-0.015464964	$+0.001682669$	-0.43964535
330	-0.11088191	-0.18905102	-0.020417779	-0.009062352	-0.43573540
Σ_1	-0.00673552	$+0.00541570$	-0.052184839	-0.048675063	-2.64060044
Σ_2	-0.00673567	$+0.00541552$	-0.052184849	-0.048675069	-2.64060046

$$\sin\varphi \cdot \tfrac{1}{2}A_1{}^{(s)} + \cos\varphi \cdot B_0{}^{(c)} = +0.00000000000029.$$

DIFFERENTIAL COEFFICIENTS.

			log coeff.
$[de/dt]_{00}$	$= -\ 2.3648522$	m'	$n\ 0.3738040$
$[d\chi/dt]_{00}$	$= +277.85744$	m'	$p\ 2.4438220$
$[di/dt]_{00}$	$= -\ 18.322835$	m'	$n\ 1.2629927$
$[d\Omega/dt]_{00}$	$= -288.76199$	m'	$n\ 2.4605400$
$[d\pi/dt]_{00}$	$= +277.35124$	m'	$p\ 2.4430301$
$[dL/dt]_{00}$	$= -927.63054$	m'	$n\ 2.9673751$

FINAL VALUES CORRESPONDING TO THE ABOVE VALUE OF m'.

$$[de/dt]_{00} = -0.00067536338$$
$$[d\chi/dt]_{00} = +0.079351564$$
$$[di/dt]_{00} = -0.0052327048$$
$$[d\Omega/dt]_{00} = -0.082465731$$
$$[d\pi/dt]_{00} = +0.079207000$$
$$[dL/dt]_{00} = -0.26491624$$

COMPARISON WITH OTHER RESULTS.

	Leverrier.	Newcomb.	Method of Gauss.
$[de/dt]_{00}$	-0.00067	-0.00067	-0.00067536
$e[d\pi/dt]_{00}$	$+0.00055$	$+0.00054$	$+0.00054202$
$[di/dt]_{00}$	-0.00523	-0.00523	-0.00523270
$\sin i\ [d\Omega/dt]_{00}$	-0.00489	-0.00488	-0.00488077
$[dL/dt]$	-0.265		-0.26491624

NOTES.

As in the previous case, the considerable disagreement of the sums of the functions near the beginning of the computation nearly disappears as the work progresses, showing that the convergence of the expansion of the perturbing function is here very rapid. The greatest error which would have arisen from the neglect of all terms from the 6th to the 11th orders would have here occurred with the coefficient $[d\chi/dt]_{00}$ and would have amounted to 1/70000th part of the remaining terms.

ACTION OF URANUS ON VENUS.

E	A	B cos ε	B sin ε	g	h
0°	369.8294733	28.016529	− 8.627090	60.35594	367.496220
30	370.1021929	30.912376	− 2.059265	3.43885	367.496155
60	370.0319613	30.136215	+ 5.076263	20.89679	367.496140
90	369.6376057	25.896024	+10.867539	95.77536	367.496165
120	369.0247981	19.327964	+13.762797	153.60480	367.496245
150	368.3577351	12.191933	+12.986248	136.75991	367.496230
180	367.8151465	6.400029	+ 8.745974	62.03086	367.496275
210	367.5424142	3.504185	+ 2.178150	3.84740	367.496245
240	367.6126214	4.280346	− 4.957378	19.92946	367.496135
270	368.0069649	8.520535	−10.748653	93.69136	367.496165
300	368.6197847	15.088597	−13.643909	150.96247	367.496240
330	369.2868724	22.224622	−12.867361	134.26735	367.496240
Σ_1	2212.9337853*	103.249680†	+ 0.356657‡	467.78032	2204.977255
Σ_2	2212.9337852	103.249675	+ 0.356658	467.78025	2204.977200

E	l	G	G′	G″	θ		
					°	′	″
0°	+1.522310	367.495771	1.6238940	0.1011370	3	55	40.93
30	+1.795095	367.496129	1.8003166	0.0051977	4	1	9.46
60	+1.724880	367.495985	1.7573874	0.0323564	4	0	5.50
90	+1.330495	367.495453	1.5044398	0.1732316	3	52	23.80
120	+0.717605	367.495105	1.0990547	0.3803068	3	38	8.91
150	+0.050560	367.495217	0.6363646	0.5847917	3	18	7.21
180	−0.492075	367.495816	0.2329563	0.7245710	2	55	22.89
210	−0.764775	367.496217	0.0134531	0.7782012	2	39	26.72
240	−0.694455	367.495988	0.0708724	0.7651843	2	43	51.72
270	−0.300145	367.495472	0.3769271	0.6763791	3	3	57.85
300	+0.312600	367.495121	0.8167025	0.5029833	3	25	59.41
330	+0.979685	367.495243	1.2686692	0.2879853	3	43	48.64
Σ_1	+3.090865	2204.973786	5.6008673	2.5065388	20	39	9.36
Σ_2	+3.090915	2204.973731	5.6001704	2.5057866	20	38	53.68

* $6a^2 + 3a^2e^2 + 6[a'^2 - 2kaa'ee' \cos K] = 2212.9337852.$

† $6[a'^2e' - kaa'e \cos K] = + 103.249685.$

‡ $- 6k'aa' \cos \varphi' \cdot e \sin K' = + 0.356657.$

Action of Uranus on Venus.

E	$\log K_0$	$\log L'$	$\log N_0$	$\log N$	$\log P$	$\log Q$
0°	0.00153256	0.27504417	0.17838936	5.725524$\overline{0}$	0.869824$\overline{4}$	3.338541$\overline{5}$
30	0.00160469	0.27514027	0.17849746	5.726567$\overline{0}$	0.871189$\overline{5}$	3.339805$\overline{6}$
60	0.00159052	0.27512139	0.17847622	5.7286906	0.8732303	3.341876$\overline{0}$
90	0.00149009	0.27498757	0.17832569	5.7313185	0.8753928	3.344187$\overline{6}$
120	0.00131281	0.27475131	0.17805993	5.7337420	0.8770917	3.3461011
150	0.00108264	0.27444453	0.17771485	5.7353151	0.8778752	3.347087$\overline{8}$
180	0.00084825	0.27413211	0.17736340	5.735623$\overline{4}$	0.877539$\overline{7}$	3.346878$\overline{9}$
210	0.00070103	0.27393587	0.17714264	5.734589$\overline{6}$	0.876182$\overline{5}$	3.345560$\overline{7}$
240	0.00074042	0.27398839	0.17720173	5.732487$\overline{0}$	0.874163$\overline{6}$	3.343532$\overline{8}$
270	0.00093336	0.27424555	0.17749102	5.7298710	0.872015$\overline{5}$	3.341311$\overline{5}$
300	0.00117044	0.27456155	0.17784648	5.7274385	0.870309$\overline{0}$	3.339439$\overline{5}$
330	0.00138191	0.27484340	0.17816353	5.7258446	0.8695041	3.338416$\overline{2}$
Σ_1	0.00719500	1.64759892	1.06733712	4.383505$\overline{4}$	5.2421585	0.056369$\overline{6}$
Σ_2	0.00719372	1.64759719	1.06733519	4.3835056	5.242159$\overline{4}$	0.0563691

E	$\log V$	J_1'	J_2	J_3	F_2
0°	3.338392$\overline{2}$	367.09942052	-0.77238411	-13.377206	$+148.77538$
30	3.3397979	366.74089875	-0.18040606	-16.560521	$+ 35.51232$
60	3.3418282	366.88453611	$+0.52246333$	-15.273725	$- 87.54088$
90	3.343931$\overline{9}$	367.40160988	$+0.87586900$	$- 9.861620$	-187.41225
120	3.3455398	367.86665960	$+0.72768333$	$- 1.774376$	-237.34142
150	3.3462248	367.95568703	$+0.32642368$	$+ 6.821041$	-223.94969
180	3.345809$\overline{7}$	367.72285452	$+0.04295564$	$+13.621489$	-150.82559
210	3.3444123	367.51508313	$+0.01342241$	$+16.804806$	$- 37.56251$
240	3.3424036	367.61069788	$+0.04551264$	$+15.518012$	$+ 85.49071$
270	3.3403134	367.89469814	-0.13535069	$+10.105905$	$+185.36209$
300	3.3386972	367.98729038	-0.54951786	$+ 2.018660$	$+235.29114$
330	3.3379911	367.66272932	-0.88324571	$- 6.576756$	$+221.89949$
Σ_1	0.0526706	2205.17145901*	$+0.01671297$	$+ 0.732854$	$- 6.15066$
Σ_2	0.0526714	2205.17070625	$+0.01671263$	$+ 0.732855$	$- 6.15055$

* $\Sigma_1(J_1' - G'') = 2202.66492021.$
$\Sigma_2(J_1' - G'') = 2202.66491965.$

ACTION OF URANUS ON VENUS.

E	F_3	$1000 \times R_0$	$1000000 \times S_0$	$1000000 \times W_0$	$1000 \times R^{(n)}$	$1000000 \times S^{(n)}$
0°	$-$ 4.0226412	0.02650922	-0.05811066	-2.918780	0.00000000	-0.08089097
30	$-$ 0.1770896	0.02652281	-0.01305234	-3.621493	$+0.01844306$	-0.01815231
60	$-$ 1.6152186	0.02667654	$+0.04940526$	-3.356828	$+0.03204873$	$+0.06853681$
90	$-$ 6.9332675	0.02691093	$+0.05269544$	-2.182313	$+0.03720412$	$+0.07285096$
120	-10.8407275	0.02710189	-0.01759673	-0.401343	$+0.03233767$	-0.02424436
150	$-$ 9.4434565	0.02715609	-0.09660941	$+1.506695$	$+0.01866093$	-0.13277471
180	$-$ 4.1342719	0.02707758	-0.10424233	$+3.017072$	0.00000000	-0.14313458
210	$-$ 0.2013093	0.02694790	-0.02527822	$+3.713882$	-0.01851786	-0.03474102
240	$-$ 1.5455370	0.02684358	$+0.07399797$	$+3.412656$	-0.03202946	$+0.10195266$
270	$-$ 6.7883563	0.02677475	$+0.10841711$	$+2.207472$	-0.03701585	$+0.14988562$
300	-10.6594103	0.02669990	$+0.05468656$	$+0.432404$	-0.03207678	$+0.07586323$
330	$-$ 9.2743255	0.02659629	-0.02803268	-1.439065	-0.01849416	-0.03898596
Σ_1	-32.8178065	0.16090871	-0.00185993	$+0.185181$	$+0.00028016$	-0.00191721
Σ_2	-32.8178047	0.16090877	-0.00186010	$+0.185178$	$+0.00028024$	-0.00191742

E	$1000 \times \left[R_0 \sin v + \left(\cos v + \cos E \right) S_0 \right]$	$1000 \times \left[-R_0 \cos v + \left(\frac{r}{a} \sec^2\varphi + 1 \right) \sin v S_0 \right]$	$1000000 \times W_0 \cos u$	$1000000 \times W_0 \sin u$	$1000 \times -2\frac{r}{a} R_0$
0°	-0.00011622	-0.02650922	-1.710234	$-$ 2.365243	-0.05265561
30	$+0.01331757$	-0.02293687	-0.357980	$-$ 3.603757	-0.05273125
60	$+0.02323048$	-0.01311517	$+1.390488$	$-$ 3.055297	-0.05317053
90	$+0.02690995$	$+0.00028955$	$+1.777154$	$-$ 1.266575	-0.05382187
120	$+0.02340803$	$+0.01365914$	$+0.399475$	$-$ 0.038680	-0.05438924
150	$+0.01366523$	$+0.02346772$	-1.372927	$-$ 0.620648	-0.05463405
180	$+0.00020848$	$+0.02707758$	-1.767867	$-$ 2.444894	-0.05452574
210	-0.01335043	$+0.02340859$	-0.392395	$-$ 3.693094	-0.05421519
240	-0.02324178	$+0.01343114$	$+1.376698$	$-$ 3.122646	-0.05387086
270	-0.02677488	-0.00003361	$+1.779939$	$-$ 1.305659	-0.05354950
300	-0.02314722	-0.01330732	$+0.429867$	$-$ 0.046772	-0.05321706
330	-0.01342562	-0.02295918	-1.315326	$-$ 0.583802	-0.05287735
Σ_1	$+0.00034177$	$+0.00123615$	$+0.118427$	-11.073532	-0.32182904
Σ_2	$+0.00034182$	$+0.00123620$	$+0.118465$	-11.073535	-0.32182921

$\sin \varphi \cdot \tfrac{1}{2}A_1^{(s)} + \cos \varphi \cdot B_0^{(c)} = + 0.000000000000020.$

DIFFERENTIAL COEFFICIENTS.

				log coeff.
$[de/dt]_{00}$	$= +$	$0.12000343''$	m'	p 9.0791936
$[d\chi/dt]_{00}$	$= +$	63.424159	m'	p 1.8022547
$[di/dt]_{00}$	$= +$	0.04158807	m'	p 8.6189687
$[d\Omega/dt]_{00}$	$= -$	65.693091	m'	n 1.8175197
$[d\pi/dt]_{00}$	$= +$	63.308999	m'	p 1.8014655
$[dL/dt]_{00}$	$= $	-113.109825	m'	n 2.0535003

FINAL VALUES CORRESPONDING TO THE ABOVE VALUE OF m'.

$$[de/dt]_{00} = +0.0000052633084''$$
$$[d\chi/dt]_{00} = +0.0027817616$$
$$[di/dt]_{00} = +0.000001824038$$
$$[d\Omega/dt]_{00} = -0.0028812762$$
$$[d\pi/dt]_{00} = +0.0027767109$$
$$[dL/dt]_{00} = -0.0049609570$$

COMPARISON WITH OTHER RESULTS.

	Leverrier.	Newcomb.	Method of Gauss.
$[de/dt]_{00}$	$+0.00000''$	$+0.00001''$	$+0.000005263''$
$e[d\pi/dt]_{00}$	$+0.00002$	$+0.00002$	$+0.000019001$
$[di/dt]_{00}$	$+0.00000$	$+0.00000$	$+0.000001824$
$\sin i \, [d\Omega/dt]_{00}$	-0.000165	-0.00017	-0.000170530

NOTES.

That a division into eight parts is here fully sufficient is shown by the agreement of the final sums. Thus the greatest effect produced by all terms from the 4th to the 7th order is seen to occur with the coefficient $[d\chi/dt]_{00}$ and to amount to but $0''.00000004$.

ACTION OF NEPTUNE ON VENUS.

E	A	B cos ε	B sin ε	g	h
0°	904.77843877	+ 9.109428	+21.528911	30.253556	904.17419
45	904.51260261	− 6.657698	+16.194152	17.117826	904.17356
90	904.39237664	−14.030230	+ 1.286186	0.107979	904.17298
135	904.48820486	− 8.689452	−14.462107	13.651972	904.17345
180	904.74393521	+ 6.236084	−21.825589	31.093114	904.17407
225	905.00974680	+22.003203	−16.490829	17.750770	904.17359
270	905.12994825	+29.375736	− 1.582864	0.163538	904.17281
315	905.03414463	+24.034972	+14.165428	13.097596	904.17334
Σ_1	3619.04469887*	+30.691018†	− 0.593356‡	61.618187	3616.69405
Σ_2	3619.04469890	+30.691025	− 0.593356	61.618164	3616.69394

E	l	G	G'	G''	θ
					° ′ ″
0°	0.53898	904.17415	0.5952295	0.0562134	1 32 17.018
45	0.27377	904.17354	0.3309885	0.0571984	1 11 14.024
90	0.15413	904.17298	0.1548950	0.0007710	0 45 6.502
135	0.24949	904.17343	0.2998555	0.0503537	1 7 39.561
180	0.50459	904.17403	0.5654480	0.0608162	1 30 28.921
225	0.77088	904.17357	0.7955805	0.0246764	1 43 33.466
270	0.89187	904.17281	0.8920680	0.0002028	1 48 0.658
315	0.79553	904.17332	0.8133615	0.0178097	1 44 14.706
Σ_1	2.08957	3616.69397	2.2076405	0.1180034	5 35 53.099
Σ_2	2.08967	3616.69386	2.2397860	0.1500382	5 46 41.757

ACTION OF NEPTUNE ON VENUS.

E	log K_0	log L_0'	log N_0	log N	log P	log Q
0°	0.00023476	0.27331427	0.17644338	5.1378654̄	9.4986215̄	2.3580297̄
45	0.00013986	0.27318776	0.17630105	5.1395214̄	9.5001506̄	2.3595431̄
90	0.00005608	0.27307605	0.17617538	5.1436917̄	9.5042639̄	2.3636152̄
135	0.00012618	0.27316951	0.17628052	5.1479186̄	9.5085363̄	2.3679232̄
180	0.00022568	0.27330217	0.17642976	5.1497409̄	9.5104807	2.3698895
225	0.00029563	0.27339543	0.17653468	5.1481063̄	9.5089743̄	2.3683773̄
270	0.00032162	0.27343007	0.17657365	5.1439578̄	9.5048849̄	2.3642800̄
315	0.00029957	0.27340068	0.17654059	5.1397097̄	9.5005900	2.3599901
Σ_1	0.00083814	1.09312256	0.70562217	0.5752556	8.0182509̄	9.4558143̄
Σ_2	0.00086124	1.09315338	0.70565684	0.5752558	8.0182511	9.4558336̄

* $4a^2 + 2a^2e^2 + 4[a'^2 - 2kaa'ee' \cos K] = 3619.04469884.$

† $4[a'^2e' - kaa'e \cos K] = + 30.691024.$

‡ $- 4k'aa' \cos \varphi' \cdot e \sin K' = - 0.593359.$

ACTION OF NEPTUNE ON VENUS.

E	$\log V$	J_1'	J_2	J_3	F_2
0°	2.3579959	902.1174996	$+0.0278169$	-43.343250	-165.19752
45	2.3595087	903.0233587	$+1.1892143$	-32.830864	-124.26237
90	2.3636147	904.1640829	$+0.1518779$	-2.911684	-9.86927
135	2.3678929	903.3068137	-1.1744691	$+28.888013$	$+110.97189$
180	2.3698530	902.1221024	-0.3409776	$+43.940408$	$+167.47400$
225	2.3683624	902.9705135	$+0.9081699$	$+33.428015$	$+126.53886$
270	2.3642798	904.1592634	$+0.1562063$	$+3.508843$	$+12.14577$
315	2.3599794	903.2945923	-0.9279474	-28.290864	-108.69539
Σ_1	9.4557434	3612.5629483*	-0.0050765	$+1.194317$	$+4.55298$
Σ_2	9.4557434	3612.5952782	-0.0050323	$+1.194300$	$+4.55299$

E	F_3	$100000 \times R_0$	$100000000 \times S_0$	$1000000 \times W_0$	$100000 \times R^{(n)}$	$1000000 \times S^{(n)}$
0°	$+0.5839042$	0.6821434	-0.4573134	-0.9883466	0.0000000	-0.006365877
45	-3.9412736	0.6867470	$+2.3281642$	-0.7513860	$+0.6746065$	$+0.032343148$
90	-0.4777116	0.6961065	$+0.3193229$	-0.0672756	$+0.9623607$	$+0.004414608$
135	$+4.0782664$	0.7008255	-2.3820088	$+0.6740551$	$+0.6818058$	-0.032772474
180	$+0.6001079$	0.7010447	-0.2565237	$+1.0297346$	0.0000000	-0.003522313
225	-4.0073139	0.7005531	$+2.5294517$	$+0.7805505$	-0.6815408	$+0.034801040$
270	-0.5873116	0.6967370	$+0.4002346$	$+0.0811602$	-0.9632323	$+0.005533205$
315	$+3.9893101$	0.6879052	-2.4698966	-0.6479491	-0.6757442	-0.034312118
Σ_1	$+0.1189889$	2.7760316	$+0.0057204$	$+0.0552726$	-0.0008716	$+0.000059623$
Σ_2	$+0.1189890$	2.7760308	$+0.0057105$	$+0.0552705$	-0.0008727	$+0.000059596$

E	$100000 \times [R_0 \sin v + (\cos v + \cos E)S_0]$	$100000 \times \left[-R_0 \cos v + \left(\frac{r}{a}\sec^2\varphi + 1 \right)\sin v S_0 \right]$	$100000 \times W_0 \cos u$	$100000 \times W_0 \sin u$	$1000 \times -2\frac{r}{a}R_0$
0°	-0.00091463	-0.68214344	-0.057911307	-0.080091019	-0.013549511
45	$+0.49123773$	-0.47994176	$+0.012282960$	-0.074127847	-0.013668483
90	$+0.69608806$	$+0.00540218$	$+0.005478547$	-0.003904555	-0.013922131
135	$+0.49653747$	$+0.49458425$	-0.066499046	-0.011017258	-0.014084333
180	$+0.00051305$	$+0.70104468$	-0.060336403	-0.083444904	-0.014116841
225	-0.49655471	$+0.49418261$	$+0.012013897$	-0.077124930	-0.014078857
270	-0.69672347	$+0.00396738$	$+0.006544151$	-0.004800407	-0.013934740
315	-0.49226067	-0.48055578	-0.064023059	-0.009971443	-0.013691533
Σ_1	-0.00103699	$+0.02827080$	-0.106225012	-0.172240885	-0.055523223
Σ_2	-0.00104018	$+0.02826932$	-0.106225248	-0.172241478	-0.055523206

$\sin\varphi \cdot \tfrac{1}{2}A_1^{(s)} + \cos\varphi \cdot B_0^{(c)} = -0.0000000000000085,$

* $\Sigma_1(J_1' - G'') = 3612.4449449.$

$\Sigma_2(J_1' - G'') = 3612.4452400,$

DIFFERENTIAL COEFFICIENTS.

log coeff.

$[de/dt]_{00} = -\overset{''}{0}.0054696734 \; m' \quad n \; 7.7379614$

$[d\chi/dt]_{00} = +21.756678 \quad\quad m' \quad p \; 1.3375926$

$[di/dt]_{00} = -0.55945727 \quad\quad m' \quad n \; 9.7477669$

$[d\Omega/dt]_{00} = -15.327159 \quad\quad m' \quad n \; 1.1854617$

$[d\pi/dt]_{00} = +21.729810 \quad\quad m' \quad p \; 1.3370559$

$[dL/dt]_{00} = -29.268164 \quad\quad m' \quad n \; 1.4663954$

FINAL VALUES CORRESPONDING TO THE ABOVE VALUE OF m'.

$[de/dt]_{00} = -\overset{''}{0}.00000027764841$

$[d\chi/dt]_{00} = +0.0011044000$

$[di/dt]_{00} = -0.000028398849$

$[d\Omega/dt]_{00} = -0.00077802855$

$[d\pi/dt]_{00} = +0.0011030360$

$[dL/dt]_{00} = -0.0014856935$

COMPARISON WITH OTHER RESULTS.

	Leverrier.	Newcomb.	Method of Gauss.
$[de/dt]_{00}$	$-\overset{''}{0}.00000$	$-\overset{''}{0}.00000$	$-\overset{''}{0}.00000028$
$e[d\pi/dt]_{00}$	$+0.00001$	$+0.00001$	$+0.00000755$
$[di/dt]_{00}$	-0.00004	-0.00003	-0.00002840
$\sin i \; [d\Omega/dt]_{00}$	-0.00006	-0.00005	-0.00004605

NOTES.

The large disagreement of the sums of the functions near the beginning of the computation is caused, as in previous cases, by the presence of the term $a'^2 e'$. The greatest disagreement in the final sums occurs in the second column and shows that the effect of all terms from the 4th to the 7th orders is to produce a change of $0''.00000003$ in the value of $[d\chi/dt]_{00}$.

EARTH.

ACTION OF MERCURY ON THE EARTH.

E	A	$B \cos \epsilon$	$B \sin \epsilon$	$1000 \times g$	h
0°	1.25770017	+0.37398164	+0.15686177	0.15586327	1.11042819
30	1.20879374	+0.24429166	+0.30606106	0.59337027	1.06575342
60	1.14343370	+0.05583076	+0.37253457	0.87910918	1.00282212
90	1.07923573	−0.14090326	+0.33847094	0.72569217	0.93669980
120	1.03345314	−0.29319554	+0.21299723	0.28738066	0.88633014
150	1.01830180	−0.36023958	+0.02973421	0.00560044	0.86904450
180	1.03773900	−0.32407097	−0.16221302	0.16667904	0.89145124
210	1.08650443	−0.19438096	−0.31141230	0.61430099	0.94482277
240	1.15158355	−0.00592005	−0.37788591	0.90454667	1.01121409
270	1.21564099	+0.19081399	−0.34382220	0.74882000	1.07228810
300	1.26156411	+0.34310630	−0.21834847	0.30200208	1.11392105
330	1.27699637	+0.41015028	−0.03508545	0.00779765	1.12755062
Σ_1	6.88547367*	+0.14973214†	−0.01605383‡	2.69558090	6.01616683
Σ_2	6.88547306	+0.14973213	−0.01605374	2.69558152	6.01615921

E	l	G	G'	G''	θ		
					°	′	″
0°	0.14093752	1.11028337	0.14207046	0.00098811	21	1	34.45
30	0.13670586	1.06515341	0.14124977	0.00394390	21	37	28.51
60	0.13427712	1.00181061	0.14149060	0.00620197	22	30	21.00
90	0.13620147	0.93572981	0.14260963	0.00543817	23	21	59.90
120	0.14078854	0.88589477	0.14348474	0.00226084	23	53	48.72
150	0.14292284	0.86903562	0.14297679	0.00004507	23	55	58.88
180	0.13995330	0.89120229	0.14152378	0.00132152	23	34	54.52
210	0.13534720	0.94401808	0.14077439	0.00462250	23	2	51.29
240	0.13403500	1.01019211	0.14138996	0.00633298	22	24	31.27
270	0.13701843	1.07154031	0.14266461	0.00489838	21	43	51.55
300	0.14130860	1.11364215	0.14347758	0.00189008	21	9	39.20
330	0.14311129	1.12754360	0.1431666$\bar{3}$	0.00004830	20	52	41.84
Σ_1	0.83130008	6.01302530	0.85343712	0.01899550	134	34	49.16
Σ_2	0.83130709	6.01302083	0.8534418$\bar{2}$	0.01899632	134	34	51.97

* $6a^2 + 3a^2e^2 + 6[a'^2 - 2kaa'ee' \cos K] = 6.8854738.$

† $6[a'^2e' - kaa'e \cos K] = + 0.14973211.$

‡ $- 6k'aa' \cos \varphi' \cdot e \sin K' = - 0.01605375.$

ACTION OF MERCURY ON THE EARTH.

E	$\log K_0$	$\log L_0'$	$\log N_0$	$\log N$	$\log P$	$\log Q$
0°	0.04527343	0.33291492	0.24335185	9.9618524	0.2031270	0.1593841
30	0.04797642	0.33646399	0.24732687	9.9917424	0.2701720	0.2100521
60	0.05211516	0.34189221	0.25340453	0.039602$\overline{0}$	0.374562$\overline{4}$	0.289540$\overline{6}$
90	0.05634203	0.34742859	0.25960067	0.0958412	0.4959354	0.3817747
120	0.05903754	0.35095528	0.26354624	0.1435571	0.5975343	0.4586143
150	0.05922392	0.35119901	0.26381890	0.163158$\overline{6}$	0.636237$\overline{4}$	0.487917$\overline{4}$
180	0.05742739	0.34884899	0.26118992	0.1459441	0.5935535	0.4565142
210	0.05475392	0.34534936	0.25727397	0.1016262	0.4927722	0.3817985
240	0.05164961	0.34128199	0.25272142	0.048225$\overline{7}$	0.375271$\overline{5}$	0.293829$\overline{0}$
270	0.04846610	0.33710662	0.24804652	0.000482$\overline{2}$	0.273610$\overline{2}$	0.2165394
300	0.04587421	0.33370403	0.24423576	9.967336$\overline{7}$	0.206076$\overline{5}$	0.164090$\overline{4}$
330	0.04461832	0.33205428	0.24238775	9.9536821	0.1814324	0.143917$\overline{9}$
Σ_1	0.31137734	2.04959742	1.51844973	0.3065179	2.350125$\overline{1}$	1.821972$\overline{5}$
Σ_2	0.31138070	2.04960185	1.51845468	0.3065326	2.3501595	1.8219999

E	$\log V$	J_1'	J_2	J_3	F_2
0°	0.1589092	0.14310992	+0.032517759	+0.009293505	−0.004694216
30	0.2080805	0.14534318	+0.064629382	+0.012521451	−0.009159128
60	0.2862522	0.14795099	+0.079427870	+0.011088166	−0.011148401
90	0.3786919	0.14823938	+0.072184328	+0.005377209	−0.010129019
120	0.4572602	0.14575314	+0.044711897	+0.003081410	−0.006374115
150	0.4878898	0.14319523	+0.004967377	−0.012020967	−0.000889820
180	0.4557266	0.14344333	−0.035694279	−0.019045640	+0.004854357
210	0.3791987	0.14602945	−0.066230415	−0.022272938	+0.009319268
240	0.2904985	0.14803624	−0.078997000	−0.020838344	+0.011308544
270	0.2141065	0.14763142	−0.071296738	−0.015126732	+0.010289159
300	0.1631852	0.14536909	−0.045398479	−0.006668768	+0.006534256
330	0.1438950	0.14322488	−0.007686344	+0.002269483	+0.001049961
Σ_1	1.8118319	0.87366271*	−0.003432232	−0.029252491	+0.000480425
Σ_2	1.8118624	0.87366354	−0.003432410	−0.029252494	+0.000480421

$* \Sigma_1(J_1' - G'') = 0.85466721.$

$\Sigma_2(J_1' - G'') = 0.85466722.$

ACTION OF MERCURY ON THE EARTH.

E	$1000 \times F_3$	R_0	S_0	W_0	$R^{(n)}$	$S^{(n)}$
0°	-0.3347370	-0.9146346	$+0.03939096$	$+0.012865148$	0.0000000	$+0.04006286$
30	-0.1102483	-0.9755942	$+0.08729250$	$+0.020012484$	-0.4949864	$+0.08857904$
60	$+0.5661785$	-1.0850638	$+0.12713088$	$+0.022775691$	-0.9476313	$+0.12820596$
90	$+1.0162060$	-1.2358930	$+0.14090497$	$+0.016043819$	-1.2358930	$+0.14090497$
120	$+0.7859945$	-1.3865198	$+0.10290744$	-0.005719611	-1.1907761	$+0.10205179$
150	$+0.1010653$	-1.4553467	$+0.01142556$	-0.036530916	-0.7172557	$+0.01126198$
180	-0.3579653	-1.3946588	-0.08289463	-0.055794474	0.0000000	-0.08152732
210	-0.1348464	-1.2530893	-0.12959913	-0.053749951	$+0.6175749$	-0.12774376
240	$+0.5468018$	-1.1065943	-0.12737525	-0.039380718	$+0.9503695$	-0.12631602
270	$+1.0072426$	-0.9942912	-0.09740907	-0.022874526	$+0.9942912$	-0.09740907
300	$+0.7898462$	-0.9252302	-0.05560180	-0.008440800	$+0.8080489$	-0.05607200
330	$+0.1166998$	-0.8987688	-0.00911126	$+0.003338197$	$+0.4560076$	-0.00924555
Σ_1	$+1.9961187$	-6.8127015	$+0.00355760$	-0.073694764	-0.3799890	$+0.00640527$
Σ_2	$+1.9961190$	-6.8129832	$+0.00350357$	-0.073760893	-0.3802614	$+0.00634761$

E	$R_0 \sin v$ $+ (\cos v + \cos E)S_0$	$-R_0 \cos v$ $+ \left(\dfrac{r}{a}\sec^2\varphi + 1\right)\sin v S_0$	$W_0 \cos(v+\pi)$	$W_0 \sin(v+\pi)$	$-2\dfrac{r}{a}R_0$
0°	$+0.0787819$	$+0.9146346$	-0.002313800	$+0.012655370$	1.7985901
30	-0.3440930	$+0.9286742$	-0.013088473	$+0.015139066$	1.9228487
60	-0.8219877	$+0.7498962$	-0.021560174	$+0.007341046$	2.1519297
90	-1.2380823	$+0.2610826$	-0.015731601	-0.003149764	2.4717860
120	-1.2947997	-0.5330552	$+0.004304192$	$+0.003766681$	2.7962929
150	-0.7369917	-1.2550382	$+0.011990880$	$+0.034506913$	2.9529690
180	$+0.1657893$	-1.3946588	-0.010034649	$+0.054884684$	2.8360980
210	$+0.8424959$	-0.9617143	-0.034466377	$+0.041244714$	2.5425788
240	$+1.0791999$	-0.3473975	-0.036894585	$+0.013770630$	2.2317474
270	$+0.9957850$	$+0.1781428$	-0.022567368	-0.003736033	1.9885826
300	$+0.7530386$	$+0.5475911$	-0.006510755	-0.005371886	1.8349430
330	$+0.4402010$	$+0.7837111$	$+0.001148460$	$+0.003134422$	1.7714297
Σ_1	-0.0399777	-0.0629896	-0.073009771	$+0.087046525$	13.6496011
Σ_2	-0.0406851	-0.0651418	-0.072714479	$+0.087139318$	13.6501948

$$\sin\varphi \cdot \tfrac{1}{2}A_1{}^{(s)} + \cos\varphi \cdot B_0{}^{(c)} = +0.00000006.$$

DIFFERENTIAL COEFFICIENTS.

				log coeff.
$[de/dt]_{00}$	$= -$	8710.1780	m'	n 3.9400270
$[d\chi/dt]_{00} = [d\pi/dt]_{00}$	$= -$	824986.23	m'	n 5.9164467
$[dp/dt]_{00}$	$= +$	18814.333	m'	p 4.2744888
$[dq/dt]_{00}$	$= -$	15740.112	m'	n 4.1970078
$[dL/dt]_{00}$	$= +$	2948201.7	m'	p 6.4695572

FINAL VALUES CORRESPONDING TO THE ABOVE VALUE OF m'.

$$[de/dt]_{00} = -0.\overset{''}{0}011613570$$
$$[d\chi/dt]_{00} = [d\pi/dt]_{00} = -0.10999815$$
$$[dp/dt]_{00} = +0.0025085775$$
$$[dq/dt]_{00} = -0.0020986812$$
$$[dL/dt]_{00} = +0.39309355$$

COMPARISON WITH OTHER RESULTS.

	Leverrier.	Newcomb.	Method of Gauss.
$[de/dt]_{00}$	$-0.\overset{''}{0}0116$	$-0.\overset{''}{0}0116$	$-0.\overset{''}{0}0116136$
$e[d\pi/dt]_{00}$	-0.00184	-0.00184	-0.00184479
$[dp/dt]_{00}$	$+0.00250$	$+0.00251$	$+0.00250858$
$[dq/dt]_{00}$	-0.00209	-0.00210	-0.00209868
$[dL/dt]_{00}$	$+0.3931$		$+0.39309355$

NOTES.

Although I and e' are here very large, the error in the approximate test with ϵ, G, G', G'' and θ is small in consequence of the smallness of the factor a'. As we approach the end of the computation, however, the difference of the sums steadily increases, indicating the rather slow convergence of the perturbing function. The greatest difference is in the coefficient $[d\pi/dt]_{00}$ where terms from the fifth to the eleventh orders inclusive amount to one sixtieth part of the remaining terms and produce an effect of $0''.0018$ in the value of $[d\pi/dt]_{00}$. A division into twelve parts is thus necessary in this case, but a comparison with the computation of the action of Mars on Mercury, and especially with the similar case of Mercury on Venus, where twenty-four points of division are employed, renders it evident that more than twelve points are in the present case unnecessary.

The agreement with previous values is exact. The results obtained by HILL in the "*New Theory*," pages 511 and 512, are,

$$[dp/dt]_{00} = +0.''0025049 \quad [dq/dt]_{00} = -0.''0020956$$

These are, however, but provisional values. (See the note to the computation of Venus on the Earth.)

ACTION OF VENUS ON THE EARTH.

E	A	log B	ε			q^2	θ'		
°			°	′	″		°	′	″
0	1.49844749	9.8537612	331	0	3.89	0.0795055$\bar{9}$	7	0	55.62
30	1.50411382	9.8546442	1	19	46.170	0.08070333	5	48	28.79
60	1.51484369	9.856764$\bar{5}$	31	34	37.20	0.08262793	4	19	13.62
90	1.52786583	9.8597314	61	38	57.20	0.08465948	3	12	16.55
120	1.5397420$\bar{1}$	9.8626537	91	29	35.875	0.08634022	2	33	43.11
150	1.54723847	9.8644870	121	8	43.24	0.0874100$\bar{9}$	2	9	29.44
180	1.54824407	9.8645724	150	43	29.14	0.08768573	1	52	5.43
210	1.54243719	9.862968$\bar{3}$	180	22	35.176	0.08700647	1	54	16.46
240	1.53142595	9.860356$\bar{8}$	210	12	5.09	0.08536569	2	37	52.10
270	1.51826327	9.857620$\bar{5}$	240	13	17.66	0.08309965	4	8	28.42
300	1.5065276$\bar{4}$	9.8554195	270	23	50.222	0.08090038	5	56	51.93
330	1.49931255	9.8540767	300	40	21.95	0.07954648	7	7	42.75
Σ_1	9.13923084*	9.1535280	1085	23	41.42	0.5024255$\bar{4}$	24	20	41.81
Σ_2	9.13923113	9.153528$\bar{1}$	905	23	41.40	0.5024255$\bar{0}$	24	20	42.41

E	1000 × r'	1000 × g	G	G'	1000000 × G''	θ		
°						°	′	″
0	2.7380635	0.002936418	0.97594422	0.522484505	5.79	47	1	40.84
30	2.3200483	0.000006753	0.98353171	0.520557558	0.04	46	40	41.61
60	1.7893068	0.003473631	0.99543723	0.519388564	6.67	46	14	52.33
90	1.3770128	0.009946420	1.00773136	0.520128988	18.98	45	55	31.27
120	1.1340363	0.013007536	1.01774388	0.521998108	24.54	45	44	23.02
150	0.9732049	0.009615238	1.02406979	0.523162060	17.95	45	37	23.13
180	0.8464697	0.003140184	1.02571566	0.522509836	5.92	45	32	21.65
210	0.8529468	0.000000056$\bar{3}$	1.02173849	0.520674148	0.00	45	32	59.47
240	1.1449685	0.003259143	1.01199593	0.519411649	6.09	45	45	34.89
270	1.7299249	0.009581104	0.99828868	0.519968638	18.53	46	11	45.77
300	2.3843853	0.012589278	0.98467983	0.521847851	24.49	46	43	7.27
330	2.7841227	0.009256115	0.97615654	0.523149586	18.07	47	3	38.83
Σ_1	10.0372301	0.038406190	6.01151675	3.127640513	73.50	277	2	0.00
Σ_2	10.0372604	0.038406192	6.01151657	3.127640978	73.57	277	2	0.08

$* \ 6a^2 + 3a^2e^2 + 6[a'^2 - 2kaa'ee'\cos K] = 9.13923110.$

ACTION OF VENUS ON THE EARTH.

E	$\log K_0$	$\log L_0'$	$\log \dot N_0$	$\log N$	$\log P$	$\log Q$
0°	0.26147483	0.60779696	0.54775634	0.2626427	0.8915846	0.820971$\overline{5}$
30	0.25683878	0.60207812	0.54149917	0.2549482	0.8714495	0.8036590
60	0.25122074	0.59513848	0.53390180	0.2468811	0.8459860	0.7827661
90	0.24707162	0.59000660	0.52828039	0.2420422	0.8253429	0.766969$\overline{7}$
120	0.24470700	0.58707937	0.52507272	0.2404872	0.8122693	0.757911$\overline{3}$
150	0.24322990	0.58524987	0.52306751	0.2402489	0.8048244	0.7529792
180	0.24217345	0.58394092	0.52163264	0.2400756	0.8019574	0.750678$\overline{7}$
210	0.24230579	0.58410492	0.52181242	0.2408209	0.8062462	0.7532935
240	0.24496050	0.58739327	0.52541673	0.2444416	0.8214721	0.7646769
270	0.25055055	0.59430993	0.53299440	0.2516543	0.8474359	0.785384$\overline{6}$
300	0.25737182	0.60273603	0.54221917	0.2600985	0.8762226	0.809011$\overline{8}$
330	0.26191246	0.60833646	0.54834645	0.264913$\overline{3}$	0.894194$\overline{8}$	0.823732$\overline{2}$
Σ_1	1.50190834	3.56408503	3.19599940	1.4946267	5.0494920	4.6860161
Σ_2	1.50190910	3.56408590	3.19600034	1.494627$\overline{8}$	5.049493$\overline{7}$	4.686018$\overline{1}$

E	$\log V$	J_1'	$1000 \times J_2$	$1000 \times J_3$	$1000 \times F_2$
0°	0.8209685	0.522862677	$-3.063273\overline{1}$	$+12.62905\overline{0}$	$+1.2372987$
30	0.8036589	0.521939851	$-0.733650\overline{4}$	$+24.87782\overline{9}$	-0.0593369
60	0.782762$\overline{8}$	0.521368465	$+2.4504401$	$+30.338229$	-1.3457284
90	0.766960$\overline{3}$	0.521723193	$+5.083579\overline{7}$	$+27.547130$	-2.2771885
120	0.757899$\overline{2}$	0.522632205	$+5.8345100$	$+17.252379$	-2.6041325
150	0.7529704	0.5231916$\overline{4}$2	$+4.4234178$	$+2.21247\overline{1}$	-2.2389562
180	0.750675$\overline{8}$	0.522862807	$+1.7386808$	$-13.54264\overline{7}$	-1.2795084
210	0.7532935	0.521968683	-0.9069972	-25.791387	$+0.017127\overline{5}$
240	0.7646739	0.521377068	$-2.685926\overline{9}$	-31.251723	$+1.3035189$
270	0.785375$\overline{3}$	0.521675619	-3.5987100	-28.460585	$+2.2349789$
300	0.808999$\overline{4}$	0.52258215$\overline{0}$	-4.0331250	-18.16587\overline{3}$	$+2.5619225$
330	0.823723$\overline{0}$	0.52318645$\overline{2}$	-4.0263352	-3.126030	$+2.196746$\overline{5}$
Σ_1	4.6859794	3.13368537$\overline{2}$*	$+0.2413060$	-2.74058\overline{5}$	-0.1266292
Σ_2	4.685981$\overline{3}$	3.133685439	$+0.2413047$	-2.740573	-0.1266288

* $\Sigma_1(J_1' - G'') = 3.13361187\overline{2}$.
$\Sigma_2(J_1' - G'') = 3.133611869$.

ACTION OF VENUS ON THE EARTH.

E	$1000 \times F_3$	R_0	$100 \times S_0$	W_0	$-\dfrac{r}{a}R_0$	$R^{(n)}$
$0°$	$+0.06535471$	-1.8283274	-1.0644424	$+0.08413477$	1.7976643	0.0000000
30	-0.00196302	-1.7898617	-0.5109526	$+0.15828193$	1.7638656	-0.9081206
60	-0.00570449	-1.7535729	$+0.5420211$	$+0.18393250$	1.7388680	-1.5314810
90	$+0.05916590$	-1.7367356	$+1.4494077$	$+0.16147390$	1.7367356	-1.7367356
120	$+0.12885901$	-1.7362036	$+1.6510300$	$+0.09963433$	1.7507629	-1.4910928
150	$+0.13426042$	-1.7386892	$+1.0760758$	$+0.01338361$	1.7639423	-0.8568990
180	$+0.06988985$	-1.7361351	$+0.1682926$	-0.07583153	1.7652520	0.0000000
210	$-0.0005974\overline{8}$	-1.7337538	-0.5029615	-0.14614352	1.7589352	$+0.8544665$
240	-0.00787443	-1.7442502	-0.6981624	-0.18183297	1.7588768	$+1.4980034$
270	$+0.05404190$	-1.7747204	-0.6225129	-0.17324740	1.7747204	$+1.7747204$
300	$+0.12215392$	-1.8154792	-0.6714245	-0.11610019	1.8002552	$+1.5855471$
330	$+0.12777090$	-1.8402326	-0.9613070	-0.01982985	1.8135046	$+0.9336770$
Σ_1	$+0.37267857$	-10.6139684	-0.0726856	-0.00606309	10.6116792	$+0.0609767$
Σ_2	$+0.3726786\overline{2}$	-10.6139933	-0.0722505	-0.00608133	10.6117037	$+0.0611087$

E	$S^{(n)}$	$\begin{array}{c}R_0\sin v\\+(\cos v+\cos E)S_0\end{array}$	$\begin{array}{c}-R_0\cos v\\+\left(\dfrac{r}{a}\sec^2\varphi+1\right)\sin vS_0\end{array}$	$W_0\cos(v+\pi)$	$W_0\sin(v+\pi)$
$0°$	-0.010825990	$-0.021288\overline{9}$	$+1.8283274$	-0.01513166	$+0.08276286$
30	-0.005184831	-0.9168211	$+1.5373039$	-0.10351881	$+0.11973727$
60	$+0.005466048$	-1.5259137	$+0.8639706$	-0.17411620	$+0.05928500$
90	$+0.014494077$	-1.7367344	-0.0001389	-0.15833153	-0.03170097
120	$+0.016373004$	-1.5075995	-0.8612810	-0.07497804	-0.06561473
150	$+0.010606704$	-0.8754611	-1.5022509	-0.00439303	-0.01264209
180	$+0.001655167$	$-0.003365\overline{9}$	-1.7361351	-0.01363832	$+0.07459502$
210	-0.004957609	$+0.8630786$	-1.5036464	-0.09371241	$+0.11214236$
240	-0.006923567	$+1.5048615$	-0.8818403	-0.17035373	$+0.06358327$
270	-0.006225129	$+1.7745751$	-0.0173138	-0.17092102	-0.02829602
300	-0.006771025	$+1.5786949$	$+0.8963895$	-0.08955312	-0.07388837
330	-0.009754749	$+0.9169364$	$+1.5955426$	-0.00682218	-0.01861936
Σ_1	-0.001026363	$+0.0253885$	$+0.1094311$	-0.53777107	$+0.14072305$
Σ_2	-0.001021537	$+0.0255735$	$+0.1094965$	-0.53769898	$+0.14062119$

$$\sin\varphi\cdot\tfrac{1}{2}A_1^{(s)}+\cos\varphi\cdot B_0^{(c)}=-0.0000000083.$$

DIFFERENTIAL COEFFICIENTS.

					log coeff.
	$[de/dt]_{00}$	$= +$	5503.0089	m'	$p\ 3.7406002$
$[d\chi/dt] =$	$[d\pi/dt]$	$= +$	1409586.4	m'	$p\ 6.1490917$
	$[dp/dt]_{00}$	$= +$	30388.832	m'	$p\ 4.4827140$
	$[dq/dt]_{00}$	$= -$	116164.73	m'	$n\ 5.0650743$
	$[dL/dt]_{00}$	$= +$	4584354.6	m'	$p\ 6.6612782$

FINAL VALUES CORRESPONDING TO THE ABOVE VALUE OF m'.

$$[d\chi/dt] = \begin{aligned} [de/dt]_{00} &= +\ 0.013483339 \\ [d\pi/dt]_{00} &= +\ 3.4537341 \\ [dp/dt]_{00} &= +\ 0.074457966 \\ [dq/dt]_{00} &= -\ 0.28462399 \\ [dL/dt]_{00} &= +11.232473 \end{aligned}$$

COMPARISON WITH OTHER RESULTS.

	Leverrier.	Newcomb.	Method of Gauss.
$[de/dt]_{00}$	$+\ 0.01344$	$+0.01348$	$+\ 0.0134833$
$e[d\pi/dt]$	$+\ 0.05796$	$+0.05792$	$+\ 0.0579231$
$[dp/dt]_{00}$	$+\ 0.07450$	$+0.07446$	$+\ 0.0744580$
$[dq/dt]_{00}$	$-\ 0.28454$	-0.28462	$-\ 0.2846240$
$[dL/dt]_{00}$	$+11.2298$		$+11.232473$

NOTES.

This computation is of special interest because, notwithstanding the low eccentricities of both the Earth and Venus, the perturbing function is but slowly convergent for this case. In 1893, the computation was effected by MR. R. T. A. INNES who employed HILL's second modification of GAUSS's method, using in the work manuscript tables prepared by himself. (See *M. N.*, Vol. LIII, No. 6. The tables were afterward published in *M. N.*, Vol. LIV, No. 5.) The values of $[dp/dt]_{00}$ and $[dq/dt]_{00}$ were also obtained by HILL in the *"New Theory,"* pages 511 and 512.

As the results of INNES differed considerably in some cases from those hitherto obtained, particularly in the case of $[de/dt]_{00}$, which agreed to the first two figures only with the values of LEVERRIER and NEWCOMB, and in the case of $[dq/dt]_{00}$, which differed in the fourth figure from the value given by HILL, and in order to make the comparison more exact, the roots in the present paper were computed by the formulas of the second method, their values being afterward verified by those of the first. It was found that the functions tabulated by MR. INNES are substantially correct, though the last two significant figures of all functions from R_0 to the end usually differ, doubtless owing to the inaccuracy of the tables employed by MR. INNES. Using the values as given by him, all of this part of his computation was duplicated, with the result that an error was found in his value of $[de/dt]_{00}$, while for $[dq/dt]_{00}$ and the other coefficients his values were found to be substantially correct. The various values here referred to are as follows:

	Innes.	Hill.
$[de/dt]_{00}$	$+ 0.013476*$	
$e[d\pi/dt]_{00}$	$+ 0.057915$	
$[dp/dt]_{00}$	$+ 0.074459$	$+0.0744329$
$[dq/dt]_{00}$	$- 0.284623$	-0.2845280
$[dL/dt]_{00}$	$+11.232490$	

It will be noticed that the results of INNES are in almost exact accordance with those here given. The disagreement of the value of $[dq/dt]_{00}$ as derived by GAUSS's method with that found by HILL is, however, a more serious matter, and is almost the sole cause of the considerable disagreement of the values of this variation in the complete perturbations of the Earth's orbit, the values of $[dq/dt]_{00}$ from the action of all of the other planets except Venus agreeing with those obtained by HILL very exactly. Using the values tabulated on page 510 of the "New Theory" and the formulas of page 511, I have duplicated the computation by HILL's methods and find the same results as he obtained. It is to be noticed that the theory of the motion of the ecliptic here given by HILL was to serve a temporary purpose only, the numerical values of the coefficients stated by LEVERRIER in the *Annales*, Vol. II, pages 94 to 96, being employed without a re-computation of them.

* The uncorrected value was $+ 0''.013156$.

ACTION OF MARS ON THE EARTH.

E	A	$B \cos \epsilon$	$B \sin \epsilon$	g	h
0°	3.12005845	−0.6857946	+1.1901000	0.028604007	2.3106194̄
22.5	3.04885416	−1.0809801	+0.7480691	0.011301685	2.3059358̄
45	3.01959381	−1.2762880	+0.1890610	0.000721878	2.3030821
67.5	3.03677529	−1.2419833	−0.4018201	0.003260792	2.3034537
90	3.09782583	−0.9832901	−0.9346186	0.017641228	2.3070230
112.5	3.19346899	−0.5395909	−1.3282200	0.035628705	2.3122967
135	3.30912609	+0.0215645	−1.5227032	0.046826348	2.3167796
157.5	3.42714623	+0.6147456	−1.4884581	0.044743825	2.3179851
180	3.52951891	+1.1496460	−1.2306997	0.030588915	2.3148501
202.5	3.60064084	+1.5448315	−0.7886687	0.012561720	2.3089820
225	3.62970226	+1.7401394	−0.2296607	0.001065206	2.3043625
247.5	3.61232195	+1.7058351	+0.3612206	0.002635147	2.3044350
270	3.55118902	+1.4471417	+0.8940191	0.016141855	2.3088270
292.5	3.45562818	+1.0034423	+1.2876203	0.033483868	2.3139342
315	3.34017005	+0.4422870	+1.4821032	0.044362582	2.3163153̄
337.5	3.22234876	−0.1508942	+1.4478582	0.042336214	2.3148421̄
Σ_1	26.59718442*	+1.8554059†	−0.1623989‡	0.185952019	18.4818589
Σ_2	26.59718440	+1.8554060	−0.1623987	0.185951956	18.4818645

* $8a^2 + 4a^2e^2 + 8[a'^2 - 2kaa'ee' \cos K] = 26.59718442$.

† $8[a'^2e' - kaa'e \cos K] = +1.8554056$.

‡ $-8k'aa' \cos \varphi' \cdot e \sin K' = -0.1623983$.

ACTION OF MARS ON THE EARTH.

E	l	G	G'	G''	θ		
					°	′	″
0°	0.789243$\bar{4}$	2.302409$\bar{1}$	0.812739$\bar{6}$	0.0152860	36	42	22.96
22.5	0.722722$\bar{7}$	2.3028298	0.7325284	0.0066997	34	27	16.75
45	0.6963160	2.3028870	0.6969608	0.0004498	33	23	4.72
67.5	0.7131259	2.3025627	0.7159948	0.0019779	33	55	44.11
90	0.7706071	2.3020189	0.7853689	0.0097577	35	54	24.65
112.5	0.8609765	2.3015508	0.889133$\bar{0}$	0.0174106	38	41	59.61
135	0.9721508	2.3014739	1.0076482	0.0201918	41	42	38.47
157.5	1.0889654	2.3019609	1.122308$\bar{7}$	0.0173190	44	30	20.12
180	1.1944730	2.3028661	1.2173683	0.0109112	46	46	8.30
202.5	1.2714631	2.3036994	1.281002$\bar{5}$	0.0042567	48	15	58.20
225	1.3051440	2.3038996	1.305961$\bar{0}$	0.0003540	48	50	43.66
247.5	1.2876912	2.3033085	1.2897047	0.0008871	48	27	7.65
270	1.222166$\bar{3}$	2.3023363	1.2343370	0.0056800	47	8	13.64
292.5	1.1214982	2.3016065	1.1465149	0.0126889	45	3	3.25
315	1.003659$\bar{1}$	2.3014626	1.0370980	0.0185863	42	25	11.06
337.5	0.887311$\bar{0}$	2.3018396	0.920298$\bar{7}$	0.0199852	39	31	20.00
Σ_1	7.953759$\bar{6}$	18.419353$\bar{5}$	8.0974817	0.0812168	332	52	47.46
Σ_2	7.9537539	18.4193582	8.0974855	0.0812251	332	52	49.69

ACTION OF MARS ON THE EARTH.

E	$\log K_0$	$\log L_0'$	$\log N_0$	$\log N$	$\log P$	$\log Q$
0°	0.14793515	0.46562072	0.39121913	9.585660$\bar{0}$	9.321168$\bar{2}$	9.6118228
22.5	0.12867513	0.44104101	0.36395436	9.5698260	9.2838199	9.5702568
45	0.12011431	0.43007010	0.35176641	9.5662159	9.2715711	9.555624$\bar{9}$
67.5	0.12442259	0.43559482	0.35790544	9.5749532	9.2853792	9.570274$\bar{3}$
90	0.14089818	0.45665643	0.38128232	9.594979$\bar{4}$	9.323744$\bar{0}$	9.612315$\bar{8}$
112.5	0.16647531	0.48915019	0.41726416	9.624091$\bar{9}$	9.382655$\bar{1}$	9.6760625
135	0.19735020	0.52805532	0.46020914	9.658890$\bar{7}$	9.455346$\bar{6}$	9.7533001
157.5	0.22941061	0.56809820	0.50425204	9.6947360	9.5321279	9.8336349
180	0.25803549	0.60355500	0.54311538	9.725999$\bar{7}$	9.600911$\bar{7}$	9.904793$\bar{6}$
202.5	0.27840311	0.62861983	0.57051120	9.7469169	9.6490816	9.954200$\bar{6}$
225	0.28660968	0.63868139	0.58149048	9.753054$\bar{3}$	9.666675$\bar{1}$	9.972014$\bar{5}$
247.5	0.28101715	0.63182707	0.57401211	9.742794$\bar{9}$	9.649583$\bar{2}$	9.954287$\bar{6}$
270	0.26293394	0.60959543	0.54972342	9.7180756	9.6011932	9.9045601
292.5	0.23607945	0.57638319	0.51334444	9.683858$\bar{5}$	9.531404$\bar{1}$	9.8327841
315	0.20515940	0.53784176	0.47098807	9.6465515	9.453399$\bar{0}$	9.752042$\bar{5}$
337.5	0.17455362	0.49936282	0.42855178	9.6122456	9.3799495	9.674968$\bar{0}$
Σ_1	1.61903635	4.27007615	3.72979435	7.249426$\bar{9}$	5.6940086	8.066447$\bar{1}$
Σ_2	1.61903696	4.27007713	3.72979553	7.249422$\bar{9}$	5.694000$\bar{4}$	8.0664686

ACTION OF MARS ON THE EARTH.

E	$\log V$	J_1'	J_2	J_3	F_2
0°	9.6084099	2.3152378	+0.11029083	+0.061812671	−0.25644059
22.5	9.5687497̄	2.3059045	+0.06946998	+0.074883922	−0.16119255
45	9.5555233̄	2.2995328	+0.01805514	+0.077080482	−0.04073852
67.5	9.5698283	2.3016313	−0.03654213	+0.068068172	+0.08658346
90	9.6101298̄	2.3103260	−0.08632858	+0.049219187	+0.20138993
112.5	9.6721978̄	2.3186977	−0.12376087	+0.023403207	+0.28620250
135	9.7488617	2.3215901	−0.14288822	−0.005449606	+0.32810925
157.5	9.8298619	2.3181659	−0.14041555	−0.032946848	+0.32073022
180	9.9024333	2.3108630	−0.11642269	−0.054902532	+0.26518896
202.5	9.9532839	2.3034774	−0.07451078	−0.067974156	+0.16994090
225	9.9719383	2.2994233	−0.02127963	−0.070171605	+0.04948689
247.5	9.9540965	2.3004768	+0.03479475	−0.061160169	−0.07783516
270	9.9033316	2.3061702	+0.08485465	−0.042311544	−0.19264158
292.5	9.8300222	2.3139376	+0.12119708	−0.016495194	−0.27745411
315	9.7479652̄	2.3199983	+0.13850882	+0.012358484	−0.31936096
337.5	9.6705453	2.3208585	+0.13455789	+0.039856619	−0.31198183
Σ_1	8.0485930	18.4831415*	−0.01520968	+0.027635537	+0.03499338
Σ_2	8.0485855̄	18.4831497	−0.01520963	+0.027635553	+0.03499343

* $\Sigma_1(J_1' - G'') = 18.4019247.$

$\Sigma_2(J_1' - G'') = 18.4019246.$

ACTION OF MARS ON THE EARTH.

E	$1000 \times F_3$	R_0	S_0	W_0	$R^{(n)}$	$S^{(n)}$
0°	− 5.019874	0.22207195	−0.00895625	+0.02403761	0.00000000	−0.00910901
22.5	− 1.341522	0.21055435	−0.00524961	+0.02748415	+0.08184381	−0.00533223
45	+ 0.173061	0.20752103	−0.00112516	+0.02773157	+0.14850060	−0.00113867
67.5	− 1.404619	0.21281662	+0.00313244	+0.02500875	+0.19788699	+0.00315267
90	− 5.191525	0.22627677	+0.00726155	+0.01896286	+0.22627677	+0.00726155
112.5	− 9.004074	0.24749832	+0.01089521	+0.00882887	+0.22720045	+0.01082573
135	−10.631982	0.27522212	+0.01347761	−0.00609013	+0.19233060	+0.01331965
157.5	− 9.129495	0.30647387	+0.01430898	−0.02537639	+0.11549298	+0.01409065
180	− 5.368216	0.33606721	+0.01279803	−0.04599728	0.00000000	+0.01258694
202.5	− 1.527816	0.35740451	+0.00883760	−0.06172287	−0.13468590	+0.00870276
225	+ 0.177177	0.36477534	+0.00302208	−0.06569865	−0.25491209	+0.00298666
247.5	− 1.210722	0.35578971	−0.00342959	−0.05556595	−0.32661056	−0.00340772
270	− 4.837361	0.33265813	−0.00898150	−0.03579916	−0.33265813	−0.00898150
292.5	− 8.543560	0.30146830	−0.01237466	−0.01405700	−0.28031952	−0.01245460
315	−10.135232	0.26942963	−0.01319021	+0.00403826	−0.19280199	−0.01334851
337.5	− 8.672134	0.24201044	−0.01181400	+0.01658570	−0.09407098	−0.01199994
Σ_1	−40.833952	2.23402218	+0.00430615	−0.07881492	−0.21326424	+0.00357711
Σ_2	−40.833942	2.23401612	+0.00430637	−0.07881474	−0.21326273	+0.00357732

$\sin \varphi \cdot \tfrac{1}{2} A_1^{(s)} + \cos \varphi \cdot B_0^{(s)} = + 0.0000000050.$

ACTION OF MARS ON THE EARTH.

E	$R_0 \sin v +$ $(\cos v + \cos E) S_0$	$-R_0 \cos v +$ $\left(\dfrac{r}{a}\sec^2 \varphi + 1\right) \sin v S_0$	$W_0 \cos (v + \pi)$	$W_0 \sin (v + \pi)$	$-2\dfrac{r}{a} R_0$
0°	−0.01791249	−0.22207195	−0.00432317	+0.02364565	−0.43669510
22.5	+0.07214538	−0.19805102	−0.01506202	+0.02298943	−0.41458381
45	+0.14689806	−0.14657925	−0.02300258	+0.01548939	−0.41012014
67.5	+0.20021147	−0.07256855	−0.02452836	+0.00487823	−0.42290136
90	+0.22612318	+0.01831801	−0.01859384	−0.00372284	−0.45255354
112.5	+0.21867465	+0.11830141	−0.00734127	−0.00490454	−0.49817351
135	+0.17313163	+0.21584084	+0.00340234	+0.00505112	−0.55697199
157.5	+0.08900258	+0.29475425	+0.00517812	+0.02484247	−0.62244500
180	−0.02559607	+0.33606721	−0.00827262	+0.04524724	−0.68340683
202.5	−0.15101812	+0.32435071	−0.03316015	+0.05205879	−0.72588458
225	−0.25917515	+0.25670929	−0.05360955	+0.03797800	−0.73820246
247.5	−0.32389097	+0.14753236	−0.05413656	+0.01252224	−0.71614631
270	−0.33246073	+0.02354205	−0.03531845	−0.00584698	−0.66531623
292.5	−0.28957300	−0.08808428	−0.01192486	−0.00744291	−0.59906678
315	−0.21131668	−0.16946344	+0.00233483	+0.00329486	−0.53246901
337.5	−0.11585773	−0.21387153	+0.00359251	+0.01619196	−0.47652121
Σ_1	−0.30030825	+0.31236276	−0.13738304	+0.12113644	−4.47573530
Σ_2	−0.30030574	+0.31236335	−0.13738259	+0.12113567	−4.47572256

DIFFERENTIAL COEFFICIENTS.

log coeff.

$$[de/dt]_{00} = - \quad 48641.893 \; m' \quad n \; 4.6870105\overline{5}$$
$$[d\chi/dt]_{00} = [d\pi/dt]_{00} = +3016769.1 \quad m' \quad p \; 6.479542\overline{1}$$
$$[dp/dt]_{00} = + \quad 19626.398 \; m' \quad p \; 4.2928406$$
$$[dq/dt]_{00} = - \quad 22258.695 \; m' \quad n \; 4.347499\overline{7}$$
$$[dL/dt]_{00} = - \quad 724628.93 \; m' \quad n \; 5.8601157$$

FINAL VALUES CORRESPONDING TO THE ABOVE VALUE OF m'.

$$[de/dt]_{00} = -0.\overset{''}{0}15723904$$
$$[d\chi/dt]_{00} = [d\pi/dt]_{00} = +0.97519611$$
$$[dp/dt]_{00} = +0.0063443986$$
$$[dq/dt]_{00} = -0.0071953108$$
$$[dL/dt]_{00} = -0.23424243$$

COMPARISON WITH OTHER RESULTS.

	Leverrier.	Innes.	Hall.	Newcomb.	Method of Gauss.
$[de/dt]_{00}$	$-0.01573''$	-0.015722	-0.0157232	$-0.01572''$	-0.0157239
$[d\pi/dt]_{00}$	$+0.9754$	$+0.975224$	$+0.9751387$	$+0.9755$	$+0.9751961$
$[dp/dt]_{00}$	$+0.00635$	$+0.0063401$	$+0.0063444$	$+0.00634$	$+0.0063444$
$[dq/dt]_{00}$	-0.00721	-0.0071898	-0.0071952	-0.00719	-0.0071953
$[dL/dt]_{00}$	-0.2337	-0.23469	-0.2342416		-0.2342424

NOTES.

In the *"New Theory,"* Page 511, HILL points out that the convergence of the expansion of the perturbing function is slow in this case, the terms of the fifth order in the inclinations and eccentricities amounting to one per cent. of those of the first order. He stated that a computation by GAUSS's method would be very desirable and consequently this was effected by DR. ASAPH HALL, JR., in July, 1891 (*A. J.* No. 244), and by INNES in November, 1891 (*M. N.*, Vol. LII, Nos. 2 and 7). HALL's computation is the first application of GAUSS's method made after the publication of HILL's memoir.

Both HALL and INNES employed the values of the elements stated by LEVERRIER; Hall divided the orbit of the Earth into twelve parts and INNES into sixteen. The values of $[dp/dt]_{00}$ and $[dq/dt]_{00}$ given by the latter were however in error owing to a misprint that occurred in HILL's original paper in the value of J_3; in *M. N.*, Vol. LII, No. 7, INNES pointed out this error but did not re-compute the variations.

The final results of the present paper were printed in *A. J.*, No. 518, but the values there given are all slightly incorrect owing to errors in some of the preliminary constants, which remained undetected even in the duplication. Upon devising new test equations these were always applied to all computations previously made and in this way the errors affecting practically every figure of the present computation were discovered. The work was then both repeated and duplicated so that it is hardly possible that any errors can yet remain in it.

The latter part of INNES' computation was also duplicated, the values of J_3, W_0, $[dp/dt]_0$ and $[dq/dt]_0$ being freed from the errors referred to by him. It is these corrected values which are given above.

It will be noticed that the agreement of the results here given with those of HALL is very exact notwithstanding the difference of the original elements used in the computation. The divergences from those of INNES are more considerable, probably because the latter computer did not employ the accurate tables of HILL.

The values obtained in the *"New Theory"* for the motion of the plane of the ecliptic are,

$$[dp/dt]_{00} = +0\overset{''}{.}0063362$$
$$[dq/dt]_{00} = -0.0072112$$

Action of Jupiter on the Earth.

E	A	B cos ε	B sin ε	g	h
0°	28.04923872	+1.4444306	+5.1077212	1.6444314	27.008467
30	27.80097551	−1.1738070	+4.481882$\bar{0}$	1.2661417	27.008061
60	27.62449781	−3.1281533	+2.6317812	0.4365772	27.007154
90	27.56719541	−3.8949449	+0.0531514	0.0001781	27.006666
120	27.6444742$\bar{4}$	−3.2687173	−2.5630644	0.4140764	27.007100
150	27.83557589	−1.4172691	−4.5158555	1.2854086	27.008066
180	28.08919180	+1.1633052	−5.2819683	1.7585429	27.008637
210	28.33731437	+3.7815432	−4.6561285	1.3665053	27.008243
240	28.5135108$\bar{5}$	+5.7358901	−2.806028$\bar{3}$	0.4963014	27.007275
270	28.57067253	+6.5026807	−0.2273984	0.0032594	27.006689
300	28.49353432	+5.8764527	+2.388817$\bar{6}$	0.3596892	27.007056
330	28.30271399	+4.0250047	+4.3416085	1.1881266	27.007962
Σ₁	168.41444773*	+7.8232080†	−0.5227409‡	5.1096185	162.045689
Σ₂	168.41444770	+7.8232076	−0.522740$\bar{6}$	5.1096197	162.045687

E	l	G	G′	G″	θ		
0°	0.977739	27.0061276	1.0387013	0.0586223	11°	36′	58.64″
30	0.729882	27.0062768	0.7909419	0.0592752	10	12	32.08
60	0.554312	27.0065429	0.5826671	0.0277442	8	38	31.93
90	0.497498	27.0066658	0.497510$\bar{9}$	0.000013$\bar{3}$	7	48	2.72
120	0.574342	27.0065199	0.600457$\bar{0}$	0.025534$\bar{7}$	8	45	10.87
150	0.764478	27.0062522	0.8240511	0.0577594	10	23	57.2$\bar{4}$
180	1.017523	27.0061314	1.0803044	0.0602760	11	50	45.30
210	1.266039	27.0062772	1.3067275	0.0387224	12	53	16.35
240	1.443204	27.0065561	1.4565395	0.0126169	13	29	4.47
270	1.500952	27.0066843	1.5010367	0.0000804	13	38	11.24
300	1.423446	27.0065354	1.4332595	0.0092925	13	21	37.74
330	1.231720	27.0062551	1.2681196	0.0346927	12	40	46.3$\bar{2}$
Σ₁	5.990566	162.0384133	6.191928$\bar{8}$	0.194086$\bar{6}$	67	42	8.95
Σ₂	5.990569	162.0384114	6.188387$\bar{7}$	0.190543$\bar{4}$	67	36	45.94

* $6a^2 + 3a^2e^2 + 6[a'^2 − 2kaa'ee' \cos K] = 168.41444773.$

† $6[a'^2e' − kaa'e \cos K] = +7.8232074.$

‡ $− 6k'aa' \cos \varphi' \cdot e \sin K' = − 0.5227409.$

ACTION OF JUPITER ON THE EARTH.

E	log K_0	log L_0'	log N_0	log N	log P	log Q
0°	0.01351621	0.29098220	0.19630730	7.850219$\overline{5}$	5.276393$\overline{7}$	6.614122$\overline{8}$
30	0.01041684	0.28686620	0.19168189	7.849083$\overline{4}$	5.271115$\overline{8}$	6.608348$\overline{4}$
60	0.00744942	0.28292144	0.18724767	7.852262$\overline{7}$	5.271354$\overline{3}$	6.607595$\overline{5}$
90	0.00606351	0.28107775	0.18517479	7.8588567	5.276992$\overline{1}$	6.6125603
120	0.00764272	0.28317853	0.18753669	7.867077$\overline{3}$	5.286497$\overline{6}$	6.622734$\overline{9}$
150	0.01081185	0.28739101	0.19227173	7.8747484	5.2973550	6.6346279
180	0.01406123	0.29170555	0.19712004	7.879861$\overline{7}$	5.3067059	6.6445510
210	0.01667410	0.29517157	0.20101379	7.881068$\overline{5}$	5.312066$\overline{1}$	6.649995$\overline{3}$
240	0.01827428	0.29729273	0.20339625	7.878019$\overline{4}$	5.311967$\overline{9}$	6.6497435
270	0.01869373	0.29784858	0.20402051	7.871485$\overline{0}$	5.306388$\overline{6}$	6.644033$\overline{0}$
300	0.01793523	0.29684340	0.20289160	7.863193$\overline{5}$	5.296800$\overline{4}$	6.6344668
330	0.01613325	0.29445437	0.20020816	7.8553923	5.2858029	6.6235786
Σ_1	0.07887909	1.74292385	1.17449955	7.190633$\overline{9}$	1.749719$\overline{6}$	9.773214$\overline{4}$
Σ_2	0.07879328	1.74280948	1.17437087	7.190634$\overline{2}$	1.7497203	9.773143$\overline{4}$

E	log V	J_1'	J_2	J_3	F_2
0°	6.612951$\overline{4}$	27.06474158$\overline{2}$	+0.24634548	+0.02063093	−6.6623169
30	6.607162$\overline{6}$	27.061450021	+0.21012324	+0.32748544	−5.8459973
60	6.607039$\overline{6}$	27.022786983	+0.12128740	+0.54805936	−3.4327952
90	6.6125600	26.992014082	+0.00316191	+0.62325071	−0.0693287
120	6.622223$\overline{3}$	27.02154083$\overline{0}$	−0.11740027	+0.53291210	+3.3431631
150	6.633472$\overline{6}$	27.060747670	−0.21222668	+0.30124976	+5.8903093
180	6.6433468	27.06639528$\overline{2}$	−0.25546365	−0.00966336	+6.8895984
210	6.6492222	27.041108427	−0.23112984	−0.31651859	+6.0732768
240	6.6494917	27.00800500$\overline{2}$	−0.14158677	−0.53709407	+3.6600761
270	6.644031$\overline{4}$	26.992057684	−0.01087596	−0.61228634	+0.2966099
300	6.6342813	27.004932957	+0.12157000	−0.52194700	−3.1158827
330	6.6228858	27.037481542	+0.21569955	−0.29028297	−5.6630276
Σ_1	9.7693340	162.188402634*	−0.02524781	+0.03289796	+0.6818428
Σ_2	9.7693345	162.184859426	−0.02524778	+0.03289801	+0.6818424

* $\Sigma_1(J_1' − G'') = 161.994316084.$

$\Sigma_2(J_1' − G'') = 161.994316076.$

ACTION OF JUPITER ON THE EARTH.

E	F_3	$1000 \times R_0$	$1000 \times S_0$	$1000 \times W_0$	$1000 \times R^{(n)}$	$1000 \times S^{(n)}$
$0°$	-0.14994500	3.5906058	-0.02485742	$+0.00562843$	0.0000000	-0.02528143
30	-0.11198359	3.5669877	-0.02409495	$+0.13045170$	$+1.8097792$	-0.02445007
60	-0.03628181	3.5812719	-0.01504685	$+0.22107406$	$+3.1277000$	-0.01517409
90	$+0.00006601$	3.6318358	-0.00001620	$+0.25540271$	$+3.6318358$	-0.00001620
120	-0.04119166	3.7069905	$+0.01547114$	$+0.22249821$	$+3.1836522$	$+0.01534248$
150	-0.12070192	3.7863426	$+0.02555514$	$+0.12714491$	$+1.8660682$	$+0.02518930$
180	-0.16035007	3.8464128	$+0.02722801$	-0.00750003	0.0000000	$+0.02677889$
210	-0.12100025	3.8690539	$+0.02153453$	-0.14361296	-1.9068325	$+0.02122624$
240	-0.04149417	3.8484336	$+0.01189802$	-0.24048156	-3.3051260	$+0.01179908$
270	$+0.00005467$	3.7925083	$+0.00121404$	-0.26976404	-3.7925083	$+0.00121404$
300	-0.03599894	3.7183323	-0.00934064	-0.22557067	-3.2473996	-0.00941964
330	-0.11169659	3.6453213	-0.01884026	-0.12397381	-1.8495240	-0.01911793
Σ_1	-0.46526165	22.2920469	$+0.00535226$	-0.02435156	-0.2411734	$+0.00404529$
Σ_2	-0.46526167	22.2920496	$+0.00535230$	-0.02435149	-0.2411816	$+0.00404538$

E	$1000 \times [R_0 \sin v + (\cos v + \cos E)S_0]$	$1000 \times \left[-R_0 \cos v + \left(\frac{r}{a}\sec^2\varphi + 1\right)\sin v S_0 \right]$	$1000 \times W_0 \cos(v+\pi)$	$1000 \times W_0 \sin(v+\pi)$	$1000 \times -2\frac{r}{a}R_0$
$0°$	-0.0497148	-3.5906058	-0.00101227	$+0.00553665$	-7.0607738
30	$+1.7678935$	-3.0981989	-0.08531744	$+0.09868428$	-7.0303590
60	$+3.1124039$	-1.7713812	-0.20927556	$+0.07125644$	-7.1024820
90	$+3.6313250$	$+0.0608776$	-0.25043240	-0.05014132	-7.2636717
120	$+3.1675401$	$+1.9264211$	-0.16743708	-0.14652740	-7.4761534
150	$+1.8214372$	$+3.3200889$	-0.04173395	-0.12010043	-7.6826719
180	-0.0544560	$+3.8464128$	-0.00134888	$+0.00737774$	-7.8218420
210	-1.9439518	$+3.3453092$	-0.09208972	$+0.11020057$	-7.8505000
240	-3.3167069	$+1.9516989$	-0.22529979	$+0.08409148$	-7.7614089
270	-3.7919952	$+0.0611766$	-0.26614166	-0.04405982	-7.5850167
300	-3.2561646	-1.7957525	-0.17399247	-0.14355750	-7.3742992
330	-1.8818157	-3.1224540	-0.04265147	-0.11640599	-7.1847533
Σ_1	-0.3970983	$+0.5667933$	-0.77836605	-0.12182259	-44.5969593
Σ_2	-0.3971070	$+0.5667994$	-0.77836664	-0.12182271	-44.5969726

$$\sin\varphi \cdot \tfrac{1}{2}A_1^{(s)} + \cos\varphi \cdot B_0^{(c)} = -0.0000000000093.$$

DIFFERENTIAL COEFFICIENTS.

				log coeff.
	$[de/dt]_{00}$	$= -$	$85.760340 \; m'$	$n \; 1.9332865$
$[d\chi/dt]_{00} = [d\pi/dt]_{00}$		$= +$	$7298.7450 \quad m'$	$p \; 3.8632482$
	$[dp/dt]_{00}$	$= -$	$26.316855 \; m'$	$n \; 1.4202340$
	$[dq/dt]_{00}$	$= -$	$168.14734 \; m'$	$n \; 2.2256900$
	$[dL/dt]_{00}$	$= -$	$9631.7202 \quad m'$	$n \; 3.9837038$

FINAL VALUES CORRESPONDING TO THE ABOVE VALUE OF m'.

$$[de/dt]_{00} = -0.081841849$$
$$[d\chi/dt]_{00} = [d\pi/dt]_{00} = +6.9652565$$
$$[dp/dt]_{00} = -0.025114405$$
$$[dq/dt]_{00} = -0.16046446$$
$$[dL/dt]_{00} = -9.1916336$$

COMPARISON WITH OTHER RESULTS.

	Leverrier.	Newcomb.	Method of Gauss.
$[de/dt]_{00}$	-0.08182	-0.08182	-0.0818418
$e[d\pi/dt]_{00}$	$+0.11679$	$+0.11677$	$+0.1168153$
$[dp/dt]_{00}$	-0.02501	-0.02511	-0.0251144
$[dq/dt]_{00}$	-0.16041	-0.16047	-0.1604644
$[dL/dt]_{00}$	-9.1916		-9.1916336

NOTES.

The very close agreement of the sums toward the end of this computation is owing to the circularity of the two orbits and to their small mutual inclination. It is evident that a division into eight parts would have been sufficient, while the errors arising from a division into only six parts would have been almost inappreciable.

In this, as in several other cases, the divergence from the last figure of NEWCOMB's results is rather larger than was to have been expected. The values stated by NEWCOMB were computed to one more significant figure than was published to insure the accuracy of the final figure given. The uncertainty of this figure is evidently due to neglected terms in the series employed by LEVERRIER and NEWCOMB. In the present case we obtain for $[de/dt]_{00}$,

Computed from the six even points of division.............$-0''.0818428$
Computed from the six odd points of division ....$-0 \;.0818409$,

and the difference between any two corresponding values for any other coefficient is even less than this.

The values of the coefficients which define the motion of the plane of the ecliptic are stated by HILL as follows:

$$[dp/dt]_{00} = -0.0251149''$$
$$[dq/dt]_{00} = -0.1604628$$

ACTION OF SATURN ON THE EARTH.

E	A	$B \cos \epsilon$	$B \sin \epsilon$	g	h
0°	92.9909218	+14.3293908	+1.6679163	0.795517	90.703702
30	92.7594069	+12.2250618	+6.1218324	10.716773	90.705031
60	92.3168471	+ 8.1696352	+8.9277816	22.792314	90.707414
90	91.7819295	+ 3.2497638	+9.3339213	24.913194	90.708424̄
120	91.2980364̄	− 1.2162762	+7.2314183	14.953684	90.707081
150	90.9947748	− 4.0318145	+3.1836393	2.898337	90.704732
180	90.9533006̄	− 4.4424293	−1.7248166	0.850721	90.703708
210	91.1846750	− 2.3380978	−6.1787321	10.916915	90.705020
240	91.6269534̄	+ 1.7173265	−8.9846857	23.083782	90.707394
270	92.1617302	+ 6.6371970	−9.3908217	25.217872	90.708453
300	92.6457640	+11.1032373	−7.2883175	15.189927	90.707143
330	92.9493070	+13.9187769	−3.2405388	3.002863	90.704751
Σ_1	551.8318232̄*	+29.6608843†	−0.1707036‡	77.665945	544.236442
Σ_2	551.8318234	+29.6608873	−0.1706996	77.665954	544.236411̄

E	l	G	G'	G''	θ		
0°	+2.001263	90.703603	2.0057342	0.0043727	8°	33′	39.757″
30	+1.768419	90.703703̄	1.8341642̄	0.0644170	8	18	56.347
60	+1.323475	90.704603	1.4944314	0.1681447	7	46	25.450
90	+0.787549̄	90.705370̄	1.0517493̄	0.2611465	6	53	59.891
120	+0.304998	90.705257	0.5874560	0.2806341	5	36	19.643
150	+0.004086	90.704380	0.1809884	0.1765509	3	35	46.006
180	−0.036365	90.703605	0.0803976	0.1166593	2	40	11.401
210	+0.193697	90.703690	0.4578844	0.2628566	5	6	24.385
240	+0.633602	90.704569̄	0.9146652̄	0.2782376	6	34	30.181
270	+1.167320	90.705348	1.3729264	0.2025015	7	33	52.413
300	+1.652663	90.705263̄	1.7502260̄	0.0956818	8	11	49.887
330	+1.958598	90.704378	1.9757281̄	0.0167564	8	31	21.103
Σ_1	+5.879636	544.226899	6.8329103	0.9437302	39	22	56.319
Σ_2	+5.879669̄	544.226868	6.8734407	0.9842289	40	0	20.145

* $6a^2 + 3a^2e^2 + 6[a'^2 - 2kaa'ee' \cos K] = 551.8318229.$

† $6[a'^2e' - kaa'e \cos K] = + 29.6608842.$

‡ $- 6k'aa' \cos \varphi' \cdot e \sin K' = - 0.1707000.$

ACTION OF SATURN ON THE EARTH.

E	$\log K_0$	$\log L_0'$	$\log N_0$	$\log N$	$\log P$	$\log Q$
0°	0.00730944	0.28273527	0.18703836	7.0561506̄	3.4235951̄	5.2855436̄
30	0.00689441	0.28218320	0.18641769	7.0572862	3.4236026	5.2857705
60	0.00602140	0.28102171	0.18511178	7.0610565̄	3.4252108̄	5.2877346̄
90	0.00473960	0.27931572	0.18319348	7.0664172	3.4279697	5.2907291
120	0.00312447	0.27716505	0.18077485	7.0719166̄	3.4311334	5.2937173
150	0.00128428	0.27471328	0.17801715	7.0761000̄	3.4338677̄	5.2956443
180	0.00070759	0.27394462	0.17715249	7.0778799̄	3.4354590̄	5.2968496
210	0.00259230	0.27645617	0.17997757	7.0767946	3.4354874	5.2978905
240	0.00430241	0.27873368	0.18253896	7.0731165	3.4339314	5.2966961
270	0.00570043	0.28059458	0.18463152	7.0677983	3.4311901	5.2938284
300	0.00669846	0.28192252	0.18612461	7.0622487̄	3.4279906̄	5.2902830̄
330	0.00724349	0.28264754	0.18693973	7.0579727	3.4252033	5.2872040̄
Σ_1	0.02816377	1.67552285	1.09874105	2.4023686̄	6.1773201	1.7508241̄
Σ_2	0.02845451	1.67591049	1.09917714	2.4023690̄	6.1773208̄	1.7510668̄

E	$\log V$	J_1'	J_2	J_3	F_2
0°	5.2855175̄	90.7004946	+0.12827478	−0.7968027	− 8.486434
30	5.2853861	90.7506534	+0.29200243	+1.2416991	−31.148158
60	5.2867312	90.7742572	+0.41637229	+2.9535293	−45.424968
90	5.2891707	90.7991434	+0.49225830	+3.8800058	−47.491413
120	5.2920417	90.8302700	+0.45906255	+3.7728791	−36.793788
150	5.2945889	90.8046167	+0.26449000	+2.6608537	−16.198502
180	5.2961519	90.8127812	−0.06165231	+0.8418949	+ 8.775944
210	5.2963206	90.9507883	−0.39625124	−1.1966086	+31.437674
240	5.2950355	90.8893132	−0.59035199	−2.9084413	+45.714495
270	5.2926201	90.7428396	−0.56521208	−3.8349184	+47.780934
300	5.2897120	90.6423828	−0.36124653	−3.7277897	+37.083286
330	5.2871040̄	90.6419564	−0.09682848	−2.6157627	+16.488008
Σ_1	1.7451898̄	544.6494990*	−0.00954121	+0.1352696	+ 0.868535
Σ_2	1.7451904̄	544.6899978	−0.00954107	+0.1352689	+ 0.868543

* $\Sigma_1(J_1' - G'') = 543.7057688.$
$\Sigma_2(J_1' - G'') = 543.7057689.$

ACTION OF SATURN ON THE EARTH.

E	F_3	$1000 \times R_0$	$100000 \times S_0$	$100000 \times W_0$	$10000 \times R^{(n)}$	$100000 \times S^{(n)}$
0°	−0.3555837	0.57362783	+0.02247623	−1.5471185	0.0000000	+0.0228596
30	−1.2681621	0.57437868	−0.26276188	+2.3619011	+2.9142195	−0.2666346
60	−1.2912829	0.57674453	−0.40344654	+5.6791610	+5.0370180	−0.4068583
90	−0.3956591	0.58127800	−0.31429014	+7.5403759	+5.8127800	−0.3142901
120	+0.5220384	0.58774791	−0.09358683	+7.4052797	+5.0477180	−0.0928086
150	+0.5361282	0.59426527	+0.08130415	+5.2579217	+2.9287882	+0.0801402
180	−0.3802586	0.59796705	+0.11726800	+1.6546186	0.0000000	+0.1153337
210	−1.3248861	0.59688219	+0.07295990	−2.4035193	−2.9416851	+0.0719154
240	−1.3648579	0.59139747	+0.07709136	−5.7742147	−5.0790610	+0.0764503
270	−0.4663692	0.58408733	+0.18082475	−7.5353293	−5.8408733	+0.1808248
300	+0.4731390	0.57795107	+0.28959440	−7.2511167	−5.0475302	+0.2920434
330	+0.5221422	0.57454813	+0.25136029	−5.0525337	−2.9150792	+0.2550648
Σ_1	−2.3968057	3.50543586	+0.00939762	+0.1666094	−0.0418552	+0.0070201
Σ_2	−2.3968061	3.50543960	+0.00939707	+0.1688164	−0.0418499	+0.0070205

E	$1000 \times [R_0 \sin v + (\cos v + \cos E) S_0]$	$1000 \times \left[-R_0 \cos v + \left(\frac{r}{a} \sec^2 \varphi + 1 \right) \sin v S_0 \right]$	$1000 \times W_0 \cos (v+\pi)$	$1000 \times W_0 \sin (v+\pi)$	$-2 \frac{r}{a} R_0$
0°	+0.00044952	−0.57362783	+0.002782496	−0.015218912	−0.0011280149
30	+0.28684099	−0.49762977	−0.015447198	+0.017867337	−0.0011320724
60	+0.49964759	−0.28807540	−0.053760700	+0.018305034	−0.0011438222
90	+0.58124893	+0.00346289	−0.073936353	−0.014803463	−0.0011625560
120	+0.50564826	+0.29959114	−0.055727122	−0.048767876	−0.0011853529
150	+0.29142597	+0.51791202	−0.017258560	−0.049666046	−0.0012057929
180	−0.00234536	+0.59796705	+0.002975836	−0.016276382	−0.0012159913
210	−0.29539381	+0.51865750	−0.015412218	+0.018443264	−0.0012111028
240	−0.50861524	+0.30174597	−0.054096837	+0.020191247	−0.0011927133
270	−0.58403554	+0.00617932	−0.074341448	−0.012307244	−0.0011681747
300	−0.50182267	−0.28668159	−0.055931013	−0.046147489	−0.0011462092
330	−0.28712395	−0.49766091	−0.017382540	−0.047441087	−0.0011324063
Σ_1	−0.00703790	+0.05091934	−0.213757340	−0.087914378	−0.0070121038
Σ_2	−0.00703741	+0.05092105	−0.213778317	−0.087907239	−0.0070121051

$$\sin \varphi \cdot \tfrac{1}{2} A_1{}^{(s)} + \cos \varphi \cdot B_0{}^{(c)} = + 0.00000000000028.$$

DIFFERENTIAL COEFFICIENTS.

				log coeff.
	$[de/dt]_{00}$	$= -$	1.5163927 m'	n 0.1808117
$[d\chi/dt]_{00} =$	$[d\pi/dt]_{00}$	$= +$	655.70924 m'	p 2.8167113
	$[dp/dt]_{00}$	$= -$	18.991017 m'	n 1.2785482
	$[dq/dt]_{00}$	$= -$	46.179399 m'	n 1.6644483
	$[dL/dt]_{00}$	$=$	-1514.4911 m'	n 3.1802667

FINAL VALUES CORRESPONDING TO THE ABOVE VALUE OF m'.

$$[de/dt]_{00} = -0.00043305713$$
$$[d\chi/dt]_{00} = [d\pi/dt]_{00} = +0.18725991$$
$$[dp/dt]_{00} = -0.0054235259$$
$$[dq/dt]_{00} = -0.013188086$$
$$[dL/dt]_{00} = -0.43251400$$

COMPARISON WITH OTHER RESULTS.

	Leverrier.	Newcomb.	Method of Gauss.
$[de/dt]_{00}$	-0.00044	-0.00043	-0.00043306
$e[d\pi/dt]_{00}$	$+0.00315$	$+0.00314$	$+0.00314056$
$[dp/dt]_{00}$	-0.00542	-0.00542	-0.00542353
$[dq/dt]_{00}$	-0.01317	-0.01318	-0.01318809
$[dL/dt]_{00}$	-0.4325		-0.43251400

NOTES.

Here, as in the previous case, the approximate tests completely fail with the angle ϵ, the roots G, G', G'', and with the functions which immediately depend upon these quantities. The close agreement of the final sums shows, however, that the expansion of the perturbing function is quite rapidly convergent for this case.

The values obtained by HILL in the "*New Theory*" are:

$$[dp/dt]_{00} = -0.0054237 \qquad [dq/dt]_{00} = -0.0131883$$

The agreement of the final results here obtained with all other values is satisfactory.

Action of Uranus on the Earth.

E	A	$B \cos \epsilon$	$B \sin \epsilon$	g	h
0°	369.9391833	+24.383407	−17.462615	247.29194	367.49698
45	370.9299673	+34.837638	− 7.159449	41.56718	367.49556
90	370.9628887	+34.937206	+ 7.512108	45.76306	367.49557
135	370.0188613	+24.623780	+17.957661	261.51163	367.49708̄
180	368.6506847̄	+ 9.938828	+18.058344	264.45220	367.49706̄
225	367.6596194	− 0.515405	+ 7.755178	48.77249	367.49561
270	367.6264169	− 0.614971	− 6.916382	38.79263	367.49553
315	368.5707253	+ 9.698453	−17.361932	244.44864	367.49688
Σ_1	1477.1791736̄*	+68.644470†	+ 1.191455‡	596.29983	1469.98514̄
Σ_2	1477.1791733	+68.644466	+ 1.191458	596.29994	1469.98513̄

E	l	G	G'	G''		θ	
°					°	′	″
0	+1.63126	367.495141	1.9739873	0.3408899	4	33	0.174
45	+2.62346	367.495250	2.6661957	0.0424235	4	55	28.953
90	+2.65637	367.495229	2.7027882	0.0460735	4	57	40.407
135	+1.71085̄	367.495130	2.0584810	0.3456945	4	38	13.695
180	+0.34269̄	367.495095	1.0379451̄	0.6933002	3	55	55.015
225	−0.64694	367.495249	0.1637761̄	0.8103506	2	56	52.565
270	−0.68006	367.495243	0.1303078	0.8100787	2	53	46.998
315	+0.26290	367.495069	0.9586081	0.6938970	3	50	28.866
Σ_1	+3.95026̄	1469.980708	5.8450284̄	1.8903423	16	20	22.594
Σ_2	+3.95027̄	1469.980698	5.8470609̄	1.8923656	16	21	4.079

Action of Uranus on the Earth.

E	$\log K_0$	$\log L_0'$	$\log N_0$	$\log N$	$\log P$	$\log Q$
0°	0.00205713	0.27574316	0.17917560	6.1388849	1.2833195	3.7524062
45	0.00241045	0.27621390	0.17970508	6.1440956	1.2897058	3.7584989̄
90	0.00244638	0.27626178	0.17975893	6.1544873̄	1.3001369̄	3.7689401̄
135	0.00213675	0.27584925	0.17929492	6.1638868̄	1.3084161	3.7775217
180	0.00153562	0.27504825	0.17839395	6.1668770̄	1.3097851	3.7792009̄
225	0.00086277	0.27415148	0.17738519	6.1617903	1.3035252	3.7729672
270	0.00083285	0.27411159	0.17734031	6.1515209̄	1.2932165̄	3.7626532̄
315	0.00146560	0.27495493	0.17828898	6.1419973	1.2848106	3.7542155
Σ_1	0.00687198	1.10116478	0.71466879	4.6117700̄	5.1864579	3.0632003̄
Σ_2	0.00687557	1.10116956	0.71467417	4.6117700̄	5.1864577	5.0632033̄

* $4a^2 + 2a^2e^2 + 4[a'^2 - 2kaa'ee' \cos K] = 1477.1791732.$

† $4[a'^2e' - kaa'e \cos K] = + 68.644468.$

‡ $+ 4k'aa' \cos \varphi' \cdot e \sin K' = + 1.191454.$

ACTION OF URANUS ON THE EARTH.

E	$\log V$	J_1'	J_2	J_3	F_2
0°	3.7519032	367.8221780	−0.84734235	+2.2088876	+301.43340
45	3.758436$\bar{3}$	367.4766488	−0.35515557	+4.6641774	+123.58384
90	3.768872$\bar{1}$	367.4892672	+0.38057614	+4.3579828	−129.67133
135	3.7770116	367.8350259	+0.86234922	+1.4696662	−309.97871
180	3.7781780	368.1745883	+0.82112849	−2.3088335	−311.71664
225	3.7717715	368.2455011	+0.34414218	−4.7641198	−133.86711
·270	3.761457$\bar{9}$	368.2514548	−0.29842856	−4.4579225	+119.38809
315	3.7531918	368.1823028	−0.79535630	−1.5696087	+299.69549
Σ_1	5.0604111	1471.7374883*	+0.05593372	−0.1998856	− 20.56648
Σ_2	5.060411$\bar{2}$	1471.7394786	+0.05597953	−0.1998849	− 20.56649

E	F_3	$1000000 \times R_0$	$1000000 \times S_0$	$1000000 \times W_0$	$1000000 \times R^{(n)}$	$1000000 \times S^{(n)}$
0°	+3.5560766	68.949325	+0.10018820	+1.2544323	0.000000	+0.10189713
45	+0.4865778	69.825413	+0.03716997	+2.6752556	+49.966575	+0.03761606
90	+0.8230258	71.523975	−0.03529126	+2.5611571	+71.523975	−0.03529126
135	+4.0393764	73.045932	−0.11453773	+0.8877061	+51.045918	−0.11319535
180	+3.8028425	73.443068	−0.14342145	−1.3776247	0.000000	−0.14105578
225	+0.5809413	72.487200	−0.06580265	−2.8156195	−50.655465	−0.06503154
270	+0.7097098	70.794418	+0.06221357	−2.5724979	−70.794418	+0.06221357
315	+3.7847601	69.352325	+0.12685775	−0.8818747	−49.628040	+0.12838021
Σ_1	+8.8916547	284.710786	−0.01631094	−0.1345332	+ 0.729557	−0.01223634
Σ_2	+8.8916556	284.710870	−0.01631266	−0.1345325	+ 0.728988	−0.01223053

E	$1000000 \times [R_0 \sin v + (\cos v + \cos E)S_0]$	$1000000 \times \left[-R_0 \cos v + \left(\dfrac{r}{a}\sec^2\varphi + 1\right)\sin v\, S_0\right]$	$1000000 \times W_0 \cos(v+\pi)$	$1000000 \times W_0 \sin(v+\pi)$	$1000 \times -2\dfrac{r}{a} R_0$
0°	+ 0.200376	−68.949325	−0.2256099	+1.2339773	−0.13558593
45	+50.011805	−48.728590	−2.2190515	+1.4942570	−0.13799470
90	+71.514500	+ 1.128956	−2.5113156	−0.5028130	−0.14304794
135	+51.201671	+52.095595	−0.4959295	−0.7362581	−0.14782435
180	+ 0.286843	+73.443068	−0.2477661	+1.3551611	−0.14934959
225	−50.554733	+51.949429	−2.2975222	+1.6276075	−0.14669365
270	−70.785500	+ 1.062876	−2.5379547	−0.4201590	−0.14158884
315	−49.442739	−48.631449	−0.5098809	−0.7195307	−0.13705976
Σ_1	+ 1.216219	+ 6.685575	−5.5226463	+1.6661664	−0.56957230
Σ_2	+ 1.216004	+ 6.684985	−5.5223841	+1.6660757	−0.56957246

$\sin\varphi \cdot \tfrac{1}{2}A_1^{(s)} + \cos\varphi \cdot B_0^{(c)} = -0.00000000000025.$

* $\Sigma_1(J_1' - G'') = 1469.8471460.$

$\Sigma_2(J_1' - G'') = 1469.8471130.$

Differential Coefficients.

$$\begin{aligned}
& & & & & \text{log coeff.}\\
& [de/dt]_{00} & = & + & 0.39395664 \; m' & p \; 9.5954484\\
[d\chi/dt]_{00} = & [d\pi/dt]_{00} & = & + & 129.13143 \qquad m' & p \; 2.1110320\\
& [dp/dt]_{00} & = & + & 0.53988815 \; m' & p \; 9.7323038\\
& [dq/dt]_{00} & = & - & 1.7895101 \; m' & n \; 0.2527342\overline{2}\\
& [dL/dt]_{00} & = & - & 184.51950 \qquad m' & n \; 2.2660422
\end{aligned}$$

Final Values Corresponding to the Above Value of m'.

$$\begin{aligned}
& [de/dt]_{00} & = & +0.\overset{''}{0}00017278801\\
[d\chi/dt]_{00} = & [d\pi/dt]_{00} & = & +0.0056636605\\
& [dp/dt]_{00} & = & +0.000023679306\\
& [dq/dt]_{00} & = & -0.000078487295\\
& [dL/dt]_{00} & = & -0.0080929604
\end{aligned}$$

Comparison with Other Results.

	Leverrier.	Newcomb.	Method of Gauss.
$[de/dt]_{00}$	$+0.\overset{''}{0}0002$	$+0.\overset{''}{0}0002$	$+0.\overset{''}{0}000172788$
$e[d\pi/dt]_{00}$	$+0.00009$	$+0.00010$	$+0.0000949860$
$[dp/dt]_{00}$	$+0.00002$	$+0.00002$	$+0.0000236793$
$[dq/dt]_{00}$	-0.00008	-0.00008	-0.0000784873
$[dL/dt]_{00}$	-0.0081		-0.0080929604

Notes.

It will be noticed that, owing to the very small mutual inclination, the approximate tests are here more exactly satisfied than even in the case of Saturn, where twelve points of division were employed. It is therefore evident that eight points are fully sufficient and that the greatest error arising from a division into only four points (which occurs with the coefficient $[d\pi/dt]_{00}$), could not amount to more than 1/20,000th of the whole.

The results obtained by Hill are:

$$[dp/dt]_{00} = +0.\overset{''}{0}000237 \qquad [dq/dt]_{00} = -0.\overset{''}{0}000785$$

exactly agreeing with those here given.

ACTION OF NEPTUNE ON THE EARTH.

E	A	$B \cos \epsilon$	$B \sin \epsilon$	g	h
0°	905.47785591	$+23.74841\overline{1}$	$+24.815277$	40.194861	904.17306
45	905.10315254	$+1.12729\overline{7}$	$+28.979673$	54.817494	$904.1730\overline{6}$
90	904.80486595	$-17.81434\overline{5}$	$+15.920217$	16.543607	904.17365
135	904.75792710	$-21.98076\overline{2}$	-6.713038	2.941510	904.17372
180	904.98963355	$-8.93131\overline{3}$	-25.661837	42.984089	904.17317
225	905.36405558	$+13.68980\overline{2}$	-29.826236	58.066960	$904.1730\overline{4}$
270	905.66206098	$+32.63145\overline{0}$	-16.766778	18.349806	$904.1736\overline{1}$
315	905.70928102	$+36.79786\overline{0}$	$+5.866476$	2.246398	904.17374
Σ_1	3620.93441639*	$+29.634203$†	-1.693121†	118.072363	$3616.6934\overline{0}$
Σ_2	3620.93441624	$+29.634196$	-1.693125	118.072362	3616.69355

E	l	G	G'	G''	θ
0°	1.23952	904.17301	$1.27445\overline{2}$	$0.03488\overline{2}$	2° 10′ 50.926″
45	$0.8648\overline{3}$	904.17299	$0.93007\overline{6}$	$0.06518\overline{6}$	1 54 4.344
90	0.56594	904.17363	0.596627	0.030667	1 30 33.476
135	0.51893	904.17372	0.525125	0.006195	1 23 20.558
180	0.75119	904.17312	0.809936	0.058696	1 46 33.999
225	$1.1257\overline{5}$	$904.1729\overline{7}$	1.180229	0.054414	2 7 3.526
270	$1.4231\overline{0}$	$904.1735\overline{9}$	1.437325	0.014120	2 17 46.322
315	1.47027	904.17374	1.471958	0.001688	2 18 49.394
Σ_1	$3.9798\overline{4}$	$3616.6933\overline{5}$	$4.11834\overline{0}$	$0.13836\overline{5}$	7 45 44.723
Σ_2	3.97977	$3616.6934\overline{2}$	$4.10738\overline{8}$	$0.12748\overline{3}$	7 43 17.822

ACTION OF NEPTUNE ON THE EARTH.

E	$\log K_0$	$\log L_0'$	$\log N_0$	$\log N$	$\log P$	$\log Q$
0°	0.00047205	0.27363061	0.17679925	$5.551378\overline{9}$	$9.912472\overline{9}$	2.7719098
45	0.00035874	0.27347955	0.17662931	$5.555572\overline{5}$	9.9164864	$2.775919\overline{0}$
90	0.00022606	0.27330267	0.17643033	$5.565826\overline{2}$	$9.926595\overline{7}$	$2.785989\overline{0}$
135	0.00019146	0.27325655	0.17637844	5.5760491	$9.936796\overline{0}$	2.7961727
180	0.00031307	0.27341867	0.17656083	5.5803398	9.9411989	2.8006208
225	0.00044509	0.27359467	0.17675882	$5.576268\overline{6}$	$9.937308\overline{0}$	2.7967497
270	0.00052334	0.27369899	0.17687617	$5.566135\overline{5}$	$9.927317\overline{3}$	$2.786753\overline{1}$
315	0.00053136	0.27370968	0.17688820	5.5557902	9.9169945	2.7764257
Σ_1	0.00153452	1.09405094	0.70666658	$2.263680\overline{3}$	$9.707584\overline{7}$	1.1452735
Σ_2	0.00152666	1.09404045	0.70665477	2.2636803	9.7075848	$1.145267\overline{1}$

* $4a^2 + 2a^2e^2 + 4[a'^2 - 2kaa'ee' \cos K] = 3620.93441628.$

† $4[a'^2e' - kaa'e \cos K] = +29.634198.$

‡ $-4k'aa' \cos \varphi'. e \sin K' = -1.693118.$

ACTION OF NEPTUNE ON THE EARTH.

E	$\log V$	J_1'	J_2	J_3	F_2
0°	2.7718889	903.9916758	$+0.58828709$	-13.723633	-190.54625
45	2.7758798	904.1719863	$+0.01502688$	$+ 7.632175$	-222.52292
90	2.7859714	903.5306465	-0.23467770	$+24.659009$	-122.24477
135	2.7961690	903.3686400	$+0.17398234$	$+27.382788$	$+ 51.54664$
180	2.8005856	904.0154903	$+0.15962945$	$+14.207956$	$+197.04666$
225	2.7967170	904.1717548	-0.46647484	$- 7.147853$	$+229.02334$
270	2.786744$\bar{6}$	903.5394237	-0.52719220	-24.174694	$+128.74518$
315	2.776424$\bar{7}$	903.3535982	$+0.26287497$	-26.898466	$- 45.04625$
Σ_1	1.145190$\bar{5}$	3615.0772363*	-0.01395336	$+ 0.968638$	$+ 13.00082$
Σ_2	1.145190$\bar{5}$	3615.0659793	-0.01459065	$+ 0.968644$	$+ 13.00081$

E	F_3	$100000 \times R_0$	$10000000 \times S_0$	$100000 \times W_0$	$100000 \times R^{(n)}$	$10000000 \times S^{(n)}$
0°	$- 5.0649535$	1.7793453	$+0.19215285$	-0.08120439	0.0000000	$+0.19543047$
45	$- 6.5852856$	1.7972262	-0.17462602	$+0.04549983$	$+1.2860826$	-0.17672179
90	$- 1.8342882$	1.8362237	-0.24659977	$+0.15062685$	$+1.8362237$	-0.24659977
135	$- 0.4453113$	1.8790643	$+0.15337661$	$+0.17125249$	$+1.3131267$	$+0.15157902$
180	$- 5.4164244$	1.9016607	$+0.27295016$	$+0.08971979$	0.0000000	$+0.26844799$
225	$- 6.9853355$	1.8852158	-0.09387083	-0.04482073	-1.3174254	-0.09277065
270	$- 2.0485736$	1.8383392	-0.21373143	-0.14796412	-1.8383392	-0.21373143
315	$- 0.3483062$	1.7940628	$+0.11988971$	-0.16075330	-1.2838186	$+0.12132856$
Σ_1	-14.3642397	7.3555689	$+0.00477181$	$+0.01117813$	-0.0021155	$+0.00354726$
Σ_2	-14.3642386	7.3555691	$+0.00476947$	$+0.01117829$	-0.0020347	$+0.00341514$

E	$100000 \times [R_0 \sin v + (\cos v + \cos E)S_0]$	$100000 \times \left[-R_0 \cos v + \left(\frac{r}{a} \sec^2 \varphi + 1\right) \sin vS_0\right]$	$100000 \times W_0 \cos (v+\pi)$	$100000 \times W_0 \sin (v+\pi)$	$100000 \times -2\frac{r}{a}R_0$
0°	$+0.0038431$	-1.7793453	$+0.01460463$	-0.07988026	$- 3.4990076$
45	$+1.2834469$	-1.2580636	-0.03774087	$+0.02541381$	$- 3.5518258$
90	$+1.8360068$	$+0.0258636$	-0.14769558	-0.02957146	$- 3.6724475$
135	$+1.3107603$	$+1.3464279$	-0.09567261	-0.14203579	$- 3.8026961$
180	-0.0054590	$+1.9016607$	$+0.01613612$	-0.08825681	$- 3.8671076$
225	-1.3159048	$+1.3499919$	-0.03657334	$+0.02590924$	$- 3.8151452$
270	-1.8380449	$+0.0351057$	-0.14597725	-0.02416657	$- 3.6766784$
315	-1.2819528	-1.2550750	-0.09294409	-0.13116027	$- 3.5455738$
Σ_1	-0.0036540	$+0.1832847$	-0.26293208	-0.22187510	-14.7152411
Σ_2	-0.0036504	$+0.1832812$	-0.26293091	-0.22187301	-14.7152409

$\sin \varphi \cdot \frac{1}{2}A_1^{(s)} + \cos \varphi \cdot B_0^{(c)} = + 0.0000000000000014.$

* $\Sigma_1(J_1' - G'') = 3614.9388718.$
$\Sigma_2(J_1' - G'') = 3614.9384968.$

DIFFERENTIAL COEFFICIENTS.

log coeff.

$$[de/dt]_{00} = -\ 0.011831221\ m' \quad n\ 8.0730296$$
$$[d\chi/dt]_{00} = [d\pi/dt]_{00} = +35.402545 \quad m' \quad p\ 1.5490345$$
$$[dp/dt]_{00} = -\ 0.71895833\ m' \quad n\ 9.8567037$$
$$[dq/dt]_{00} = -\ 0.85200049\ m' \quad n\ 9.930439\bar{9}$$
$$[dL/dt]_{00} = -47.671428 \quad m' \quad n\ 1.6782582$$

FINAL VALUES CORRESPONDING TO THE ABOVE VALUE OF m'.

$$[de/dt]_{00} = -0.00000060056972$$
$$[d\chi/dt]_{00} = [d\pi/dt]_{00} = +0.0017970838$$
$$[dp/dt]_{00} = -0.000036495344$$
$$[dq/dt]_{00} = -0.000043248757$$
$$[dL/dt]_{00} = -0.0024198698$$

COMPARISON WITH OTHER RESULTS.

	Leverrier.	Newcomb.	Method of Gauss.
$[de/dt]_{00}$	0.00000	0.00000	−0.00000060057
$e[d\pi/dt]_{00}$	+0.00003	+0.00003	+0.00003013915
$[dp/dt]_{00}$	−0.00004	−0.00004	−0.00003649534
$[dq/dt]_{00}$	−0.00004	−0.00004	−0.00004324876

NOTES.

The mutual inclination is here nearly twice as great as in the case of Uranus, and yet the convergence of the perturbing function is more rapid because the eccentricity of Neptune is so much smaller than that of Uranus. Hence, although the sums of ϵ, G, G', G'', etc., are in great disagreement, the final sums from which the differential coefficients are obtained are almost identical. The greatest error arising from a division into only four parts occurs with the coefficient $[dp/dt]_{00}$ and amounts to but $0''.000000000002$

The results of HILL are:

$$[dp/dt]_{00} = -0.0000366 \qquad [dq/dt]_{00} = -0.0000435$$

MARS.

ACTION OF MERCURY ON MARS.

E	A	$B \sin \epsilon$	$B \cos \epsilon$	$1000 \times g$	h
0°	2.01395126	−0.51051059	−0.07779114	1.6508943	1.87197670
30	2.19162773	−0.49427471	+0.22547307	1.5475600	2.0479547$\bar{9}$
60	2.44453125	−0.33152771	+0.47957033	0.6962245	2.29696077
90	2.71228886	+0.06587754	+0.61641571	0.0274906	2.56269605
120	2.92685114	+0.23149508	+0.59934132	0.3394636	2.77862654
150	3.02703005	+0.48090911	+0.43292228	1.4649936	2.8820729$\bar{5}$
180	2.97859089	+0.61553465	+0.16175058	2.4000193	2.83609064
210	2.79081655	+0.59929870	−0.14151362	2.2750785	2.6479592$\bar{5}$
240	2.51771727	+0.43655177	−0.39561101	1.2072061	2.3717668$\bar{1}$
270	2.23986183	+0.17090155	−0.53245622	0.1850128	2.09074126
300	2.03539750·	−0.12647099	−0.51538178	0.1013191	1.88662892
330	1.95541421	−0.37588517	−0.34896288	0.8949940	1.81063180
Σ_1	14.91703931*	+0.31507221†	+0.25187830‡	6.9351269	14.0420503$\bar{8}$
Σ_2	14.91703923	+0.31507194	+0.25187834	6.9351295	14.0420560$\bar{9}$

E	l	G	G'	G''	θ		
					°	′	″
0°	0.13564010	1.87146851	0.14234545	0.00619717	16	20	8.69
30	0.1373384$\bar{9}$	2.04755912	0.1430188$\bar{1}$	0.00528466	15	35	30.16
60	0.14123602	2.29682015	0.1434891$\bar{8}$	0.00211253	14	34	33.48
90	0.14325835	2.56269162	0.14333762	0.00007484	13	41	0.67
120	0.14189014	2.77858020	0.14279207	0.0008555$\bar{9}$	13	8	25.31
150	0.1386226$\bar{5}$	2.88188764	0.14237834	0.00357038	12	59	49.56
180	0.13616579	2.83577714	0.14242175	0.00594246	13	12	30.22
210	0.1365228$\bar{5}$	2.64761705	0.1428791$\bar{7}$	0.0060141$\bar{3}$	13	42	7.60
240	0.1396160$\bar{1}$	2.37153873	0.14339402	0.00354993	14	24	9.49
270	0.14278611	2.09069583	0.1434484$\bar{4}$	0.00061690	15	12	58.81
300	0.14243412	1.88659813	0.1428408$\bar{9}$	0.0003759$\bar{8}$	15	59	29.51
330	0.13844795	1.81033610	0.14221981	0.00347617	16	27	51.5$\bar{4}$
Σ_1	0.8369821$\bar{8}$	14.04078286	0.85728335	0.0190336$\bar{6}$	87	39	16.70
Σ_2	0.8369763$\bar{9}$	14.04078736	0.8572821$\bar{8}$	0.01903708	87	39	18.3$\bar{4}$

* $6a^2 + 3a^2e^2 + 6[a'^2 - 2kaa'ee' \cos K] = 14.91703924.$

† $- 6k'aa' \cos \varphi' \cdot e \sin K' = + 0.31507212.$

‡ $6[a'^2e' - kaa'e \cos K] = + 0.25187831.$

ACTION OF MERCURY ON MARS.

E	$\log K_0$	$\log L_0'$	$\log N_0$	$\log N$	$\log P$	$\log Q$
0°	0.02698253	0.30881667	0.21633345	$0.080204\overline{3}$	9.8417843	0.0229194
30	0.02453876	0.30558607	0.21270775	$0.031542\overline{3}$	9.7124165	9.9318941
60	0.02139874	0.30143126	0.20804354	9.9863199	$9.564698\overline{6}$	$9.832837\overline{2}$
90	0.01882471	0.29802213	0.20421541	9.9544525	$9.435056\overline{7}$	9.7499589
120	0.01734193	0.29605696	0.20200830	9.9396879	$9.347831\overline{5}$	9.6977395
150	0.01696133	0.29555239	0.20144156	$9.942799\overline{2}$	$9.317922\overline{1}$	$9.684026\overline{0}$
180	0.01752419	0.29629857	0.20227967	$9.963297\overline{0}$	9.3524331	9.7119954
210	0.01887658	0.29809086	0.20429260	9.9992756	$9.449685\overline{3}$	9.7797276
240	0.02088535	0.30075154	0.20728036	$0.045646\overline{7}$	9.5950386	9.8772472
270	0.02335208	0.30401640	0.21094579	0.0914150	9.7545935	$9.981941\overline{9}$
300	0.02583676	0.30730232	0.21463400	$0.119398\overline{5}$	9.8751689	0.0582665
330	0.02741722	0.30939106	0.21697800	0.1150661	$9.907272\overline{6}$	0.0734518
Σ_1	0.12996950	1.81065732	1.25057932	0.1345541	$7.576954\overline{9}$	$9.201005\overline{2}$
Σ_2	0.12997068	1.81065891	1.25058111	0.1345506	7.5769465	9.2010002

E	$\log V$	J_1'	J_2	J_3	F_2
0°	0.0211422	0.148654076	-0.10722959	-0.015479687	$+0.015330048$
30	9.9305073	0.148321182	-0.10335718	-0.012048361	$+0.014842503$
60	$9.832342\overline{1}$	0.145601872	-0.06932482	-0.006505466	$+0.009955397$
90	9.7499432	0.143412552	-0.01386724	-0.000353153	$+0.001978227$
120	$9.697573\overline{5}$	0.143648964	$+0.04858051$	$+0.004751607$	-0.006951532
150	$9.683358\overline{1}$	0.145958760	$+0.10133761$	$+0.007449454$	-0.014441148
180	$9.710866\overline{3}$	0.148399366	$+0.13001805$	$+0.007034457$	-0.018483798
210	9.7785043	0.148942861	$+0.12663289$	$+0.003626261$	-0.017996253
240	$9.876441\overline{5}$	0.146960494	$+0.09193479$	-0.001870362	-0.013109151
270	9.9817831	0.144077345	-0.03534552	-0.007999542	-0.005131979
300	0.0581594	0.143355345	-0.02759166	-0.013127432	$+0.003797779$
330	0.0724209	0.145910972	0.07970449	-0.015871553	$+0.011287401$
Σ_1	9.1965248	0.876620117*	$+0.06638728$	-0.025196883	-0.009461257
Σ_2	9.1965168	0.876623672	$+0.06638711$	-0.025196894	-0.009461249

* $\Sigma_1(J_1' - G'') = 0.857586462.$
$\Sigma_2(J_1' - G'') = 0.857586592.$

ACTION OF MERCURY ON MARS.

E	$1000 \times F_2$	R_0	S_0	W_0	$R^{(n)}$	$S^{(n)}$
0°	+0.3749543	−1.1968194	−0.10192940	−0.015991438	0.00000000	−0.07377742
30	+0.9451901	−1.0712026	−0.08041942	−0.009779348	−0.38240323	−0.05741705
60	+0.8478648	−0.9676664	−0.04346887	−0.004110828	−0.57689880	−0.02992414
90	+0.1645035	−0.9003960	−0.00725842	−0.000153771	−0.59093068̄	−0.00476371
120	−0.4045215	−0.8699380	+0.02266377	+0.002278066	−0.47241848	+0.01421151
150	−0.2449197	−0.8749740	+0.04587694	+0.003542283	−0.26566469	+0.02785885
180	+0.5450969	−0.9160775	+0.06265312	+0.003737622	0.00000000	+0.03761136
210	+1.2365773	−0.9949389	+0.07097316	+0.002525788	+0.30208776	+0.04309857
240	+1.1824210	−1.1083446	+0.06401101	−0.000941853	+0.60188486	+0.04013864
270	+0.4525829	−1.2337308	+0.03097697	−0.007413721	+0.80969859̄	+0.02033021
300	−0.2401087	−1.3158841	−0.02869644	−0.015188691	+0.78449755	−0.01975474
330	−0.2482279	−1.2994033	−0.08505077	−0.018952193	+0.46386745	−0.06072369
Σ_1	+2.3057068	−6.3747320	−0.02476681	−0.030217122	+0.33706513	−0.03149479
Σ_2	+2.3057062	6.3746456	−0.02490154	−0.030230962	+0.33665520	−0.03161682

E	$R_0 \sin v$ $+(\cos v + \cos E)S_0$	$-R_0 \cos v$ $+\left(\frac{r}{a}\sec^2\varphi + 1\right)\sin vS_0$	$W_0 \cos u$	$W_0 \sin u$	$-2\frac{r}{a}R_0$
0°	−0.2038588	+1.1968194	−0.004111709	+0.015453801	2.1703888̄
30	−0.7173754	+0.8165787	−0.007231890	+0.006582964	1.9693580̄
60	−0.9154636	+0.3357086	−0.004043881	+0.000738872	1.8450804
90	−0.8957943	−0.0984951	−0.000144266	−0.000053225	1.8007921
120	−0.7408612	−0.4547261	+0.001481632	+0.001730419	1.8210138
150	−0.4834776	−0.7324563	+0.000768358	+0.003457946	1.8912954
180	−0.1253062	−0.9160795	−0.000961015	+0.003611962	2.0030408
210	+0.3338218	−0.9514367	−0.001700736	+0.001867379	2.150595̄1̄
240	+0.8448003	−0.7366597	+0.000887111	−0.000316419	2.3200626
270	+1.2254640	−0.1770222	+0.007311027	+0.001229687	2.467461̄7̄
300	+1.1635310	+0.6123066	+0.011609114	+0.009794123	2.5090379
330	+0.5585553	+1.1811278	+0.005822238	+0.018035720	2.3888945
Σ_1	+0.0228415	+0.0373693	+0.004861252	+0.031012758	12.6686243̄
Σ_2	+0.0211938	+0.0382962	+0.004824731	+0.031120471	12.6683967̄

$$\sin\varphi \cdot \tfrac{1}{2}A_1^{(s)} + \cos\varphi \cdot B_0^{(c)} = +0''.000000008.$$

DIFFERENTIAL COEFFICIENTS.

					log coeff.
$[de/dt]_{00}$	$= +$	2517.5250	m'	p	3.4009738
$[d\chi/dt]_{00}$	$= +$	46380.761	m'	p	4.6663379̄
$[di/dt]_{00}$	$= +$	558.61256	m'	p	2.7471107
$[d\Omega/dt]_{00}$	$= +$	110961.28	m'	p	5.0451714
$[d\pi/dt]_{00}$	$= +$	46438.628	m'	p	4.6668794
$[dL/dt]_{00}$	$= +$	1455134.1	m'	p	6.1629030

FINAL VALUES CORRESPONDING TO THE ABOVE VALUE OF m'.

$$[de/dt]_{00} = +0.00033567000$$
$$[d\chi/dt]_{00} = +0.0061841007$$
$$[di/dt]_{00} = +0.000074481672$$
$$[d\Omega/dt]_{00} = +0.014794833$$
$$[d\pi/dt]_{00} = +0.0061918174$$
$$[dL/dt]_{00} = +0.19401785$$

COMPARISON WITH OTHER RESULTS.

	Leverrier.	Newcomb.	Method of Gauss.
$[de/dt]_{00}$	+0.00036	+0.00033	+0.0003357
$e[d\pi/dt]_{00}$	+0.00058	+0.00057	+0.0005775
$[di/dt]_{00}$	+0.00008	+0.00007	+0.0000745
$\sin i\, [d\Omega/dt]_{00}$	+0.00047	+0.00048	+0.0004778

NOTES.

On account of the large eccentricities of both orbits and the high mutual inclination, the coefficients of the expansion diminish but slowly. Thus the combined effect of all terms from the 6th to the 11th orders is 1/30th of the whole with $[de/dt]_{00}$, 1/90th with $[d\pi/dt]_{00}$, and 1/200th with $[di/dt]_{00}$. Yet all of the variations are very small on account of the smallness of the mass of Mercury. A comparison with the computation of Mars on Mercury renders it evident that a division into twelve parts is sufficient and that terms of orders above the eleventh are wholly inappreciable.

ACTION OF VENUS ON MARS.

E	A	$B \cos \epsilon$	$B \sin \epsilon$	$1000 \times g$	h
0°	2.41946745	−0.9101348	−0.4038413	0.003995803	1.8967820
30	2.47732532	−0.5530947	−0.8457642	0.017525935	1.9549495
60	2.63301427	−0.0217107	−1.0499345	0.027008907	2.1104659
90	2.85220984	+0.5416339	−0.9616453	0.022657521	2.3292431
120	3.07987480	+0.9859909	−0.6045536	0.008954722	2.5566759̄
150	3.25131035	+1.1922952	−0.0743414	0.000135408	2.7282304
180	3.31318850	+1.1052676	+0.4869209	0.005808973	2.7904410
210	3.24523275	+0.7482272	+0.9288436	0.021138190	2.7227529
240	3.06934808	+0.2168441	+1.1330145	0.031452377	2.5468467
270	2.84005464	−0.3465004	+1.0447250	0.026741552	2.3172324
300	2.62248756	−0.7908574	+0.6876330	0.011585000	2.0993294̄
330	2.47124772	−0.9971620	+0.1574211	0.000607167	1.9481304
Σ_1	17.13738066*	+0.5853997†	+0.2492390‡	0.088805782	14.0005408
Σ_2	17.13738062	+0.5854002	+0.2492388	0.088805773	14.0005387

E	l	G	G'	$10000 \times G''$	θ		
					°	′	″
0°	0.5226610̄	1.8967805	0.5226665	0.0403053	31	39	49.77
30	0.5223513	1.9549433̄	0.5223748̄	0.1716188	31	7	34.86
60	0.5225239	2.1104579̄	0.5225564	0.2449047	29	50	30.35
90	0.5229423	2.3292377	0.5229663	0.1860051	28	17	2.93
120	0.5231745̄	2.5566741	0.5231829	0.0669458	26	53	44.43
150	0.5230555	2.7282304	0.5230555̄	0.0000489	25	58	2.26
180	0.5227230	2.7904401	0.5227279̄	0.0398246	25	38	46.94
210	0.5224554	2.7227494	0.5224737	0.1485921	25	58	48.56
240	0.5224760̄	2.5468406	0.5225067	0.2363523	26	55	59.83
270	0.5227977	2.3172260	0.5228262	0.2207297	28	21	35.72
300	0.5231337	2.0993259̄	0.5231478̄	0.1054853	29	56	49.75
330	0.5230928	1.9481302	0.5230936	0.0059582̄	31	12	36.97
Σ_1	3.1366929̄	14.0005190	3.1367881	0.7338180	170	55	41.07
Σ_2	3.1366950	14.0005170̄	3.1367900	0.7338528̄	170	55	41.30

* $6a^2 + 3a^2e^2 + 6[a'^2 - 2kaa'ee' \cos K] = 17.13738065.$

† $6[a'^2e' - kaa'e \cos K] = +0.5854002.$

‡ $-6k'aa' \cos \varphi' \cdot e \sin K' = +0.2492389.$

ACTION OF VENUS ON MARS.

E	log K_0	log L_0'	log N_0	log N	log P	log Q
0°	0.10710693	0.41334627	0.33316554	$0.153729\overline{2}$	$0.011039\overline{6}$	$0.208876\overline{8}$
30	0.10323050	0.40834940	0.32760279	$0.142061\overline{0}$	$9.968134\overline{5}$	0.1785258
60	0.09431348	0.39683240	0.31477271	0.1149515	9.8630204	$0.105342\overline{5}$
90	0.08414014	0.38365383	0.30007678	$0.082005\overline{6}$	$9.731224\overline{9}$	0.0148651
120	0.07564158	0.37216263	0.28775224	$0.052407\overline{9}$	$9.609667\overline{5}$	9.9324836
150	0.07024983	0.36559243	0.27991042	0.0325881	9.5264184	$9.876617\overline{5}$
180	0.06843912	0.36323215	0.27727293	0.0260737	$9.497959\overline{3}$	9.8576733
210	0.07032296	0.36568773	0.28001691	$0.033967\overline{9}$	$9.529635\overline{6}$	$9.878974\overline{8}$
240	0.07586487	0.37290310	0.28807661	0.0551373	9.6160289	$9.937208\overline{2}$
270	0.08461922	0.38427537	0.30077024	$0.085851\overline{8}$	$9.740182\overline{2}$	0.0216495
300	0.09502686	0.39775494	0.31580088	$0.119114\overline{4}$	9.8727052	0.1128332
330	0.10383003	0.40912260	0.32846370	0.1449403	9.9748268	$0.183786\overline{0}$
Σ_1	0.51639284	2.31668149	1.81684091	$0.521413\overline{9}$	$8.470420\overline{8}$	$0.154417\overline{5}$
Σ_2	0.51639268	2.31668136	1.81684084	0.5214145	8.4704222	$0.154418\overline{6}$

E	log V	J_1'	J_2	J_3	F_2
0°	0.2088756	0.52281610	−0.0030557479	−0.012760938	+0.0014450479
30	0.1785212	0.52261080	−0.0058421642	−0.016264386	+0.0030263615
60	$0.105336\overline{4}$	0.52273704	−0.0069459413	−0.015093265	$+0.003756935\overline{3}$
90	0.0148609	0.52303324	−0.0063117580	−0.009561780	+0.0034410135
120	9.9324822	0.52319006	−0.0041023613	−0.001152260	+0.0021632475
150	9.8766174	$0.523079\overline{8}1$	−0.0007324438	+0.007882145	+0.0002660128
180	9.8576725	0.52281605	+0.0030484592	+0.015121070	−0.0017423280
210	$9.878971\overline{9}$	0.52262308	+0.0062562314	+0.018625039	−0.0033236408
240	$9.937203\overline{3}$	$0.5226684\overline{7}$	+0.0079298727	+0.017454980	−0.0040542164
270	0.0216445	0.52292962	+0.0074476695	+0.011924014	−0.0037382940
300	0.1128306	0.52316808	+0.0048298806	+0.003513972	−0.0024605273
330	0.1837858	$0.523119\overline{2}8$	+0.0008867096	−0.005521489	−0.0005632929
Σ_1	0.1544005	$3.1373958\overline{0}$*	+0.0017041620	+0.007083559	−0.0008918410
Σ_2	$0.154401\overline{7}$	3.13739582	+0.0017042445	+0.007083543	−0.0008918399

* $\Sigma_1(J_1' - G'') = 3.1373224\overline{2}$.

$\Sigma_2(J_1' - G'') = 3.13732243$.

ACTION OF VENUS ON MARS.

E	$100000 \times F_2$	R_0	$1000 \times S_0$	$1000 \times W_0$	$R^{(n)}$	$S^{(n)}$
0°	− 2.705324	−1.4244790	−3.460778	−20.67006	0.0000000	−0.002504942
30	− 0.803003	−1.3866027	−6.000157	−24.54092	−0.4949963	−0.004283931
60	+ 5.500746	−1.3028006	−6.111878	−19.19609	−0.7766973	−0.004207440
90	+ 9.816975	−1.2077652	−4.678320	− 9.84176	−0.7926572	−0.003070385
120	+ 7.633806	−1.1282518	−2.631099	− 0.95528	−0.6126954	−0.001649853
150	+ 0.880725	−1.0779056	−0.461908	+ 5.93579	−0.3272800	−0.000280495
180	− 3.932916	−1.0618130	+1.648220	+10.88332	0.0000000	+0.000989445
210	− 2.161955	−1.0812437	+3.609363	+14.08781	+0.3282934	+0.002191792
240	+ 4.374565	−1.1352348	+5.187600	+15.12313	+0.6164876	+0.003252928
270	+ 9.225323	−1.2184712	+5.773047	+12.58406	+0.7996835	+0.003788855
300	+ 7.735212	−1.3155488	+4.427365	+ 4.61417	+0.7842976	+0.003047815
330	+ 1.648024	−1.3961376	+0.822269	− 8.41473	+0.4984001	+0.000587075
Σ_1	+18.606089	−7.3681280	−0.940570	−10.20081	+0.0113925	−0.001072047
Σ_2	+18.606089	−7.3681260	−0.935706	−10.18975	+0.0114435	−0.001067089

E	$R_0 \sin v$ $+(\cos v + \cos E)S_0$	$-R_0 \cos v$ $+\left(\frac{r}{a}\sec^2\varphi +1\right)\sin v S_0$	$W_0 \cos u$	$W_0 \sin u$	$-2\frac{r}{a}R_0$
0°	−0.0069216	+1.4244790	−0.005314674	+0.019975127	2.5832417
30	−0.7611745	+1.1593980	−0.018148167	+0.016519714	2.5492068
60	−1.1839523	+0.5449664	−0.018883467	+0.003450266	2.4840909
90	−1.2020643	−0.1220026	−0.009233404	−0.003406539	2.4155306
120	−0.9266826	−0.6439879	−0.000621304	−0.000725630	2.3617340
150	−0.4956901	−0.9571936	+0.001287534	+0.005794466	2.3299411
180	−0.0032964	−1.0618130	−0.002798312	+0.010517416	2.3216920
210	+0.4917082	−0.9631868	−0.009486007	+0.010415471	2.3371562
240	+0.9297081	−0.6522758	−0.014244152	+0.005080676	2.3763511
270	+1.2126214	−0.1251906	−0.012409750	−0.002087273	2.4369424
300	+1.1939212	+0.5533939	−0.003526734	−0.002975358	2.5083988
330	+0.7575012	+1.1728178	+0.002585060	+0.008007817	2.5667361
Σ_1	+0.0027764	+0.1647626	−0.045388643	+0.035322497	14.6355085
Σ_2	+0.0029019	+0.1646422	−0.045404734	+0.035243656	14.6355132

$$\sin\varphi \cdot \tfrac{1}{2}A_1^{(s)} + \cos\varphi \cdot B_0^{(c)} = 0.''0000000073.$$

DIFFERENTIAL COEFFICIENTS.

$$[de/dt]_{00} = + \quad\;\; 324.6318 \; m' \quad p \; 2.5113911$$
$$[d\chi/dt]_{00} = + \; 201915.56 \quad m' \quad p \; 5.3051698$$
$$[di/dt]_{00} = - \quad\;\; 5236.2608 \; m' \quad n \; 3.7190213$$
$$[d\Omega/dt]_{00} = + \; 126021.28 \quad m' \quad p \; 5.1004439$$
$$[d\pi/dt]_{00} = + \; 201981.28 \quad m' \quad p \; 5.3053112$$
$$[dL/dt]_{00} = +1681713.6 \quad m' \quad p \; 6.2257520$$

FINAL VALUES CORRESPONDING TO THE ABOVE VALUE OF m'.

$$[de/dt]_{00} = +0.0007954049$$
$$[d\chi/dt]_{00} = +0.49472856$$
$$[di/dt]_{00} = -0.012829757$$
$$[d\Omega/dt]_{00} = +0.30877426$$
$$[d\pi/dt]_{00} = +0.49488961$$
$$[dL/dt]_{00} = +4.1204933$$

COMPARISON WITH OTHER RESULTS.

	Leverrier.	Newcomb.	Method of Gauss.
$[de/dt]_{00}$	+0.00080	+0.00079	+0.000795
$e[d\pi/dt]_{00}$	+0.04618	+0.04614	+0.0461574
$[di/dt]_{00}$	−0.01280	−0.01284	−0.012830
$\sin i \; [d\Omega/dt]_{00}$	+0.00993	+0.00998	+0.009972
$[dL/dt]_{00}$	+4.117		+4.120493

NOTES.

The very close agreement of the sums of the functions near the beginning of the computation is caused by the great circularity of the orbit of Venus. The discrepancies increase however as the work proceeds because of the high eccentricity of Mars and the rather large mutual inclination. All terms from the 6th to the 11th orders, inclusive, produce an effect equal to 1/30th of the whole in the very small coefficient $[de/dt]_{00}$, and 1/1000th of the whole in $[d\Omega/dt]_{00}$. Yet it is evident that terms of the twelfth and higher orders are wholly inappreciable.

ACTION OF THE EARTH ON MARS.

E	A	$B \cos \epsilon$	$B \sin \epsilon$	$1000 \times g$	h
0°	2.88085183	−0.8153552	−1.1018750	0.34149937	1.8833414
30	2.95824702	−0.0872040	−1.3962265	0.54832354	1.9597848
60	3.13279096	+0.6917421	−1.2860904	0.46523043	2.133265$\overline{1}$
90	3.36510700	+1.3127653	−0.8009781	0.18045398	2.3654511
120	3.59664230	+1.6094626	−0.0708746	0.00141289	2.5975998
150	3.76166099	+1.5023341	+0.7085891	0.14122581	2.7632686
180	3.80855449	+1.0200854	+1.3285566	0.49646115	2.8103218
210	3.72106141	+0.2919341	+1.6229077	0.74082034	2.7223893
240	3.52632175	−0.4870119	+1.5127725	0.64368353	2.526938$\overline{6}$
270	3.28390784	−1.1080347	+1.0276597	0.29704612	2.284191$\overline{0}$
300	3.06247042	−1.4047322	+0.2975561	0.02490366	2.0633533
330	2.91764743	−1.2976042	−0.4819076	0.06532098	1.9197321
Σ_1	20.00763175*	+0.6141908†	+0.6800452‡	1.97319103	14.0148199
Σ_2	20.00763169	+0.6141906	+0.6800443	1.97319077	14.014816$\overline{9}$

E	l	G	G'	G''	θ		
					°	′	″
0	0.9972292	1.8831367	0.9976157	0.00018178	46	42	32.02
30	0.9981810	1.9594937	0.9987522	0.00028018	45	33	35.1$\overline{5}$
60	0.999244$\overline{7}$	2.1330727	0.9996552	0.00021818	43	12	19.69
90	0.999374$\overline{7}$	2.365395$\overline{3}$	0.9995068	0.00007633	40	32	44.90
120	0.9987612	2.597599$\overline{5}$	0.9987622	0.00000054	38	19	18.09
150	0.9981111	2.763239$\overline{7}$	0.9981913	0.00005120	36	56	40.43
180	0.9979514	2.8102243	0.998225$\overline{9}$	0.00017698	36	35	10.16
210	0.9983909	2.7222314	0.9988211	0.00027246	37	17	7.3$\overline{7}$
240	0.999102$\overline{0}$	2.5267718	0.9995236	0.00025487	38	58	33.58
270	0.9994356	2.2840897	0.999667$\overline{1}$	0.00013009	41	25	15.69
300	0.9988359	2.0633420	0.9988593	0.00001208	44	5	20.00
330	0.9976341	1.9196952	0.9977050	0.00003411	46	7	50.4$\overline{5}$
Σ_1	5.9911243	14.014147$\overline{0}$	5.992641$\overline{9}$	0.00084443	247	53	13.54
Σ_2	5.991127$\overline{4}$	14.0141449	5.992643$\overline{5}$	0.00084437	247	53	13.9$\overline{8}$

* $6a^2 + 3a^2e^2 + 6[a'^2 - 2kaa'ee' \cos K] = 20.00763172.$

† $6[a'^2e' - kaa'e \cos K] = +0.6141907.$

‡ $-6k'aa' \cos \varphi' \cdot e \sin K' = +0.6800448.$

ACTION OF THE EARTH ON MARS.

E	$\log K_0$	$\log L_0'$	$\log N_0$	$\log N$	$\log P$	$\log Q$
0°	0.25724275	0.60257672	0.54204483	0.3085062	0.3612353	0.5756272
30	0.24243067	0.58425966	0.52198206	0.279659$\overline{0}$	0.2795065	0.509435$\overline{0}$
60	0.21406633	0.54897788	0.48324184	0.2277018	0.1185796	0.3818936
90	0.18497596	0.51250382	0.44306081	0.172790$\overline{7}$	9.937459$\overline{0}$	0.2419337
120	0.16284593	0.48455413	0.41218099	0.129268$\overline{5}$	9.784678$\overline{0}$	0.126877$\overline{2}$
150	0.15007482	0.46834273	0.39423491	0.1040949	9.6895845	0.056903$\overline{3}$
180	0.14686278	0.46425584	0.38970669	0.0998549	9.666574$\overline{2}$	0.0407933
210	0.15317152	0.47227918	0.39859494	0.116878$\overline{7}$	9.719220$\overline{9}$	0.0805051
240	0.16915889	0.49254542	0.42101792	0.1535252	9.8408510	0.1719333
270	0.19422097	0.52412777	0.45588060	0.2048055	0.0114576	0.301948$\overline{3}$
300	0.22440852	0.56187400	0.49741669	0.2597585	0.1924850	0.442601$\overline{5}$
330	0.24970710	0.59326698	0.53185208	0.300384$\overline{5}$	0.327171$\overline{5}$	0.5489966
Σ_1	1.17458520	3.15478399	2.74560896	1.178615$\overline{1}$	9.9644030	1.7397260
Σ_2	1.17458104	3.15478014	2.74560540	1.1786131	9.964399$\overline{9}$	1.739721$\overline{9}$

E	$\log V$	J_1'	J_2	J_3	F_2
0°	0.5755790	0.99892684	−0.018227992	+0.028809050	+0.018467466
30	0.5093633	0.99952648	−0.022898545	+0.020515873	+0.023400805
60	0.3818419	0.99990323	−0.021379458	+0.006120062	+0.021554920
90	0.241917$\overline{3}$	0.99967015	−0.013761084	−0.010518960	+0.013424123
120	0.126877$\overline{1}$	0.99911770	−0.001690224	−0.024941756	+0.001187862
150	0.0568937	0.99877639	+0.011675876	−0.033284760	−0.011875978
180	0.0407608	0.99892204	+0.022549333	−0.033314577	−0.022266656
210	0.0804536	0.99942130	+0.027739796	−0.025024224	−0.027199994
240	0.171881$\overline{7}$	0.99985593	+0.025696938	−0.010634071	−0.025354120
270	0.3019193	0.99982014	+0.017055202	+0.006002126	−0.017223615
300	0.4425985	0.99929729	+0.004464507	+0.020427743	−0.004987052
330	0.5489877	0.99880864	−0.008398230	+0.028776417	+0.008076789
Σ_1	1.7395389	5.99602303*	+0.011413104	−0.013533549	−0.011397580
Σ_2	1.739534$\overline{9}$	5.99602310	+0.011413015	−0.013533528	−0.011397570

* $\Sigma_1(J_1' - G'') = 5.99517860.$
$\Sigma_2(J_1' - G'') = 5.99517873.$

ACTION OF THE EARTH ON MARS.

E	$1000 \times F_3$	R_0	S_0	W_0	$R^{(n)}$	$S^{(n)}$
0°	−0.13911285	−2.0302096	−0.026171976	+0.10810004	0.0000000	−0.018943511
30	−0.51397488	−1.9019956	−0.029451014	+0.06531256	−0.6789837	−0.021027135
60	−0.65316258	−1.6889696	−0.023181583	+0.01388518	−1.0069217	−0.015958290
90	−0.40694037	−1.4884239	−0.012395858	−0.01871310	−0.9768538	−0.008135411
120	−0.02612649	−1.3462165	−0.001540204	−0.03342035	−0.7310608	−0.000965798
150	+0.08995598	−1.2702099	+0.007499119	−0.03789964	−0.3856685	+0.004553852
180	−0.20223795	−1.2578222	+0.014435140	−0.03668660	0.0000000	+0.008665574
210	−0.63957154	−1.3082364	+0.019136251	−0.03045222	+0.3972143	+0.011620521
240	−0.80757787	−1.4237326	+0.020598419	−0.01635704	+0.7731558	+0.012916412
270	−0.54879816	−1.6023546	+0.016496265	+0.01146539	+1.0516264	+0.010826511
300	−0.11741630	−1.8174946	+0.004601712	+0.05641742	+1.0835451	+0.003167836
330	+0.07369507	−1.9931947	−0.012572900	+0.10202141	+0.7115406	−0.008976672
Σ_1	−1.94563404	−9.5644451	−0.011258492	+0.09193865	+0.1187184	−0.011117777
Σ_2	−1.94563390	−9.5644151	−0.011288137	+0.09173440	+0.1188753	−0.011138334

E	$R_0 \sin v$ $+(\cos v + \cos E)S_0$	$- R_0 \cos v$ $+ \left(\dfrac{r}{a}\sec^2\varphi + 1\right)\sin v S_0$	$W_0 \cos u$	$W_0 \sin u$	$-2\dfrac{r}{a}R_0$
0°	−0.0523440	+2.0302096	+0.027794619	−0.10446568	3.6817114
30	−1.0803151	+1.5681913	+0.048299056	−0.04396512	3.4967331
60	−1.5490313	+0.6794308	+0.013659050	−0.00249569	3.2204119
90	−1.4807799	−0.1636143	−0.017556368	−0.00647718	2.9768478
120	−1.1074128	−0.7656902	−0.021736270	−0.02538610	2.8179922
150	−0.5982290	−1.1202178	−0.008220822	−0.03699731	2.7456152
180	−0.0288703	−1.2578222	+0.009432839	−0.03545319	2.7502731
210	+0.5690362	−1.1796147	+0.020504958	−0.02251409	2.8278111
240	+1.1509410	−0.8419068	+0.015406346	−0.00549521	2.9802545
270	+1.5938315	−0.1824413	−0.011306572	−0.00190172	3.2047092
300	+1.6480563	+0.7672282	−0.043121307	−0.03637964	3.4654744
330	+1.0579845	+1.6887208	−0.031341647	−0.09708793	3.6643987
Σ_1	+0.0613389	+0.6114494	+0.001435277	−0.20967551	18.9161075
Σ_2	+0.0615274	+0.6110240	+0.000378605	−0.20894335	18.9161151

$$\sin\varphi \cdot \tfrac{1}{2}A_1^{(s)} + \cos\varphi \cdot B_0^{(c)} = + 0.000000102.$$

DIFFERENTIAL COEFFICIENTS.

			log coeff.
$[de/dt]_{00}$	$= +$	7024.3393 m'	p 3.8466055
$[d\chi/dt]_{00}$	$= +$	749340.69 m'	p 5.8746793
$[di/dt]_{00}$	$= +$	104.61082 m'	p 2.0195766
$[d\Omega/dt]_{00}$	$= -$	747594.66 m'	n 5.8736662
$[d\pi/dt]_{00}$	$= +$	748950.76 m'	p 5.8744532
$[dL/dt]_{00}$	$= +$	2175235.9 m'	p 6.3375064

FINAL VALUES CORRESPONDING TO THE ABOVE VALUE OF m'.

$$[de/dt]_{00} = +0.021481158$$
$$[d\chi/dt]_{00} = +2.2915614$$
$$[di/dt]_{00} = +0.00031991074$$
$$[d\Omega/dt]_{00} = -2.2862242$$
$$[d\pi/dt]_{00} = +2.2903688$$
$$[dL/dt]_{00} = +6.6520970$$

COMPARISON WITH OTHER RESULTS.

	Leverrier.	Newcomb.	Method of Gauss.
$[de/dt]_{00}$	$+0.02151$	$+0.02148$	$+0.02148116$
$e[d\pi/dt]_{00}$	$+0.21276$	$+0.21374$	$+0.21361818$
$[di/dt]_{00}$	$+0.00030$	$+0.00032$	$+0.00031991$
$\sin i \, [d\Omega/dt]_{00}$	-0.07391	-0.07379	-0.07383093
$[dL/dt]_{00}$	$+6.638$		$+6.6520970$

NOTES.

As in all cases in which Mars is the disturbed body, the gradual increase in the discrepancies in the sums of the functions as the computation proceeds is caused principally by the large value of e. The greatest effect which is here produced by the inclusion of all terms from the fifth to the eleventh orders occurs with the coefficient $[d\pi/dt]_{00}$ and amounts to $0''.0007$. It is evident that a division into twelve parts is fully sufficient.

ACTION OF JUPITER ON MARS.

E	A	$B \cos \epsilon$	$B \sin \epsilon$	g	h
0°	29.52014024	+6.924444	−4.477639	1.2637451	27.008366
30	29.73057269	+8.555992	−0.736229	0.0341655	27.006182
60	29.83402815	+8.090126	+3.325863	0.6972206	27.007669̄
90	29.81017790	+5.651672	+6.620206	2.7625134	27.011679
120	29.66910846	+1.894010	+8.264081	4.3047762	27.014415
150	29.44492347	−2.175995̄	+7.817013	3.8516159	27.013253
180	29.19030067	−5.467791	+5.398794	1.8371936	27.009343
210	28.96977017	−7.099343	+1.657384	0.1731441	27.006389
240	28.84611897	−6.633477	−2.404708	0.3644905	27.006977̄
270	28.85987156	−4.195019	−5.699050	2.0472288	27.010168
300	29.01103882	−0.437358̄	−7.342925	3.3985953	27.012567
330	29.25541959	+3.632647	−6.895857	2.9973538	27.011662
Σ_1	176.07073531*	+4.369955̄†	+2.763466‡	11.8660213	162.059336
Σ_2	176.07073538	+4.369955̄	+2.763467	11.8660215	162.059333

E	l	G	G'	G''	θ		
					°	′	″
0°	2.448742	27.0064605	2.4695960	0.0189481	17	39	53.68
30	2.661358	27.0061300	2.6618860	0.0004753	18	17	56.75
60	2.763328̄	27.0066036	2.7737001	0.0093077	18	43	15.01
90	2.735467	27.0074648	2.7765210	0.0368400	18	48	58.35
120	2.591662	27.0078870	2.6581522	0.0599626	18	28	29.57
150	2.368639	27.0074649	2.4330416̄	0.0586151	17	39	45.34
180	2.117926	27.0066097	2.1522663	0.0316074	16	30	39.82
210	1.900349	27.0061336	1.9039718̄	0.0033673	15	24	39.18
240	1.776111̄	27.0064416	1.7842098̄	0.0075644	14	55	27.23
270	1.786671	27.0071624	1.8310752	0.0413982	15	15	16.64
300	1.935440̄	27.0075479	2.0032755	0.0628165	16	2	14.95
330	2.180725	27.0071916	2.2348561	0.0496603	16	53	32.61
Σ_1	13.633208	162.0415503	13.8411999̄	0.1902067	102	20	0.26
Σ_2	13.633209	162.0415473	13.8413516	0.1903562	102	20	8.87

* $6a^2 + 3a^2e^2 + 6[a'^2 - 2kaa'ee' \cos K] = 176.07073528$.

† $6[a'^2e' - kaa'e \cos K] = +4.369954$.

‡ $-6k'aa' \cos \varphi' \cdot e \sin K' = +2.763466$.

ACTION OF JUPITER ON MARS.

E	$\log K_0$	$\log L_0{}'$	$\log N_0$	$\log N$	$\log P$	$\log Q$
0°	0.03165483	0.31498612	0.22325507	8.3476455$\bar{6}$	5.7990877$\bar{2}$	7.1391284
30	0.03402451	0.31811150	0.22676030	8.3623557$\bar{0}$	5.8175286$\bar{6}$	7.1576473
60	0.03565117	0.32025551	0.22916442	8.3954324	5.8524487	7.1929772
90	0.03602468	0.32074766	0.22971622	8.4366023	5.8931984$\bar{4}$	7.2342427
120	0.03469756	0.31899875	0.22775524	8.4742980$\bar{0}$	5.9283892	7.2695995$\bar{5}$
150	0.03164635	0.31497492	0.22324250	8.4991683$\bar{3}$	5.9492925$\bar{5}$	7.2899854
180	0.02757619	0.30960108	0.21721367	8.5057537	5.9513987	7.2909894$\bar{4}$
210	0.02396325	0.30482489	0.21185335	8.4928484$\bar{4}$	5.9346402$\bar{2}$	7.2731852$\bar{2}$
240	0.02244999	0.30282273	0.20960576	8.4633475	5.9029922	7.2413643$\bar{3}$
270	0.02347170	0.30417466	0.21112344	8.4239469$\bar{9}$	5.8638333	7.2029262
300	0.02598821	0.30750252	0.21485868	8.3844577	5.8269721	7.1668224$\bar{4}$
330	0.02889111	0.31133796	0.21916251	8.3560126$\bar{6}$	5.8027960	7.1428978
Σ_1	0.17801795	1.87416671	1.32185284	0.5709348	5.2612881$\bar{1}$	3.3008810
Σ_2	0.17802160	1.87417159	1.32185832	0.5709355$\bar{3}$	5.2612888	3.3008846$\bar{6}$

E	$\log V$	$J_1{}'$	J_2	J_3	F_2
0°	7.1387522	27.0207890	−0.20899481	+0.30968500	+ 5.840160
30	7.1576379	27.0065721	−0.03630747	−0.02674093	+ 0.960260
60	7.1927926	27.0099958	+0.15284461	−0.36413185	− 4.337906
90	7.2335125	27.0290264	+0.31348417	−0.61207993	− 8.634698
120	7.2684108	27.0492502	+0.40064241	−0.70414548	−10.778797
150	7.2888221	27.0526590	+0.38540712	−0.61566134	−10.195688
180	7.2903609$\bar{9}$	27.0334483	+0.26823852	−0.37034175	− 7.041617
210	7.2731181$\bar{1}$	27.0094582	+0.08088102	−0.03392190	− 2.161717
240	7.2412135	27.0106120	−0.12278062	+0.30345694	+ 3.136448
270	7.2021019	27.0363209	−0.28345023	+0.55139880	+ 7.433241
300	7.1655732$\bar{2}$	27.0521350	−0.35632057	+0.64347029	+ 9.577337
330	7.1419111	27.0423420	−0.32638726	+0.55499853	+ 8.994231
Σ_1	3.2971031	162.1762303*	+0.13362954	−0.18200685	− 3.604375
Σ_2	3.2971036$\bar{6}$	162.1763786	+0.13362735	−0.18200677	− 3.604371

* $\Sigma_1(J_1{}' - G'') = 161.9860236.$
$\Sigma_2(J_1{}' - G'') = 161.9860224.$

Action of Jupiter on Mars.

E	F_3	R_0	$1000 \times S_0$	$1000 \times W_0$	$1000 \times R^{(n)}$	$1000 \times S^{(n)}$
0°	−0.11464084	0.011523729	+0.08005004	+0.4190397	0.000000	+0.05794093
30	−0.02209150	0.011964482	+0.01088811	−0.0398940	+4.271140	+0.00777378
60	+0.08535418	0.012921829	−0.07058125	−0.5615348	+7.703673	−0.04858841
90	+0.09095508	0.014185469	−0.13852471	−1.0407895	+9.309934	−0.09091387
120	−0.02126155	0.015433331	−0.17072189	−1.3081943	+8.381048	−0.10705259
150	−0.14774803	0.016299442	−0.15776374	−1.2103390	+4.948932	−0.09580229
180	−0.16666131	0.016504164	−0.10615120	−0.7376107	0.000000	−0.06372374
210	−0.05846157	0.015978601	−0.03427562	−0.0686500	−4.851517	−0.02081393
240	+0.07437981	0.014896574	+0.03689409	+0.5347724	−8.089562	+0.02313475
270	+0.10831696	0.013593080	+0.09184683	+0.8860645	−8.921148	+0.06027915
300	+0.01978452	0.012429971	+0.12131684	+0.9434385	−7.410440	+0.08351498
330	−0.09401607	0.011688531	+0.11863615	+0.7635196	−4.172629	+0.08470265
Σ_1	−0.12304519	0.083709598	−0.10919337	−0.7100892	+0.584719	−0.05477408
Σ_2	−0.12304513	0.083709605	−0.10919298	−0.7100884	+0.584712	−0.05477451

E	$R_0 \sin v + (\cos v + \cos E)S_0$	$-R_0 \cos v + \left(\dfrac{r}{a}\sec^2 \varphi + 1\right)\sin v S_0$	$1000 \times W_0 \cos u$	$1000 \times W_0 \sin u$	$-2\dfrac{r}{a}R_0$
0°	+0.000160100	−0.011523729	+0.10774325	−0.4049515	−0.020897865
30	+0.006498115	−0.010046690	−0.02950189	+0.0268547	−0.021996162
60	+0.011621455	−0.005638033	−0.55239000	+0.1009292	−0.024638457
90	+0.014136555	+0.001045999	−0.97645422	−0.3602495	−0.028370938
120	+0.012896599	+0.008458998	−0.85083667	−0.9937043	−0.032306104
150	+0.007784436	+0.014315488	−0.26253500	−1.1815227	−0.035231972
180	+0.000212302	+0.016504164	+0.18965406	−0.7128120	−0.036086942
210	−0.007299886	+0.014215603	+0.04622538	−0.0507547	−0.034538456
240	−0.012311629	+0.008381401	−0.50369057	+0.1796589	−0.031182525
270	−0.013542396	+0.001084104	−0.87379071	−0.1469683	−0.027186159
300	−0.011129593	−0.005518211	−0.72109467	−0.6083574	−0.023700620
330	−0.006127613	−0.009949901	−0.23455823	−0.7265980	−0.021488839
Σ_1	+0.001449234	+0.010664590	−2.33061460	−2.4392371	−0.168812513
Σ_2	+0.001449211	+0.010664603	−2.33061467	−2.4392385	−0.168812526

$$\sin \varphi \cdot \tfrac{1}{2}A_1^{(s)} + \cos \varphi \cdot B_0^{(c)} = -0.00000000019.$$

DIFFERENTIAL COEFFICIENTS.

				log coeff.
$[de/dt]_{00}$	= +	165.70584	m'	p 2.2193378
$[d\chi/dt]_{00}$	= +	13074.175	m'	p 4.1164143
$[di/dt]_{00}$	= −	268.82366	m'	n 2.4294675
$[d\Omega/dt]_{00}$	= −	8712.2760	m'	n 3.9401316
$[d\pi/dt]_{00}$	= +	13069.631	m'	p 4.1162634̄
$[dL/dt]_{00}$	= −	19334.282	m'	n 4.2863281̄

FINAL VALUES CORRESPONDING TO THE ABOVE VALUE OF m'.

$$[de/dt]_{00} = + 0.15813453$$
$$[d\chi/dt]_{00} \doteq +12.476799$$
$$[di/dt]_{00} = - 0.25654077$$
$$[d\Omega/dt]_{00} = - 8.3142000$$
$$[d\pi/dt]_{00} = +12.472464$$
$$[dL/dt]_{00} = -18.450874$$

COMPARISON WITH OTHER RESULTS.

	Leverrier.	Newcomb.	Method of Gauss.
$[de/dt]_{00}$	+ 0.15810	+0.15818	+ 0.1581345
$e[d\pi/dt]_{00}$	+ 1.16323	+1.16372	+ 1.1632822
$[di/dt]_{00}$	− 0.25648	−0.25655	− 0.2565408
$\sin i \, [d\Omega/dt]_{00}$	− 0.26864	−0.26850	− 0.2684974
$[dL/dt]_{00}$	−18.450		−18.450874

NOTES.

The very exact agreement of the final sums shows that for this case the expansion of the perturbing function is highly convergent. The greatest effect of all terms from the sixth to the eleventh orders inclusive occurs with $[de/dt]_{00}$, and amounts to but 1/100000th of the whole variation.

This computation has been twice effected by DR. ARTHUR B. TURNER from the same elements as are here employed. His first computation was made by HILL's first method, exactly as here, and was presented as a *Thesis to the Faculty of the*

Graduate School of the University of Pennsylvania, 1902.   The values of the functions in this computation agree practically throughout with those here given, but the last two figures usually differ because eight place logarithms are here employed in certain parts of the work.

DR. TURNER'S second computation, (*A. N.*, 3065), was made according to the method developed by DR. LOUIS ARNDT[30].   The two papers taken together are of high value since they afford a means of comparing the labor and accuracy appertaining to the two very different methods.   It is DR. TURNER'S opinion that while the formulas of ARNDT'S method are presented in a more symmetric form yet they are less accurate in application than those of DR. HILL.   This is confirmed by the circumstance that the residual from the equation $[da/dt]_{00} = 0$ is 300 times larger with the former method than with the latter.

DR. TURNER'S results from ARNDT'S method, which agree almost exactly with his earlier values and with those here obtained, are as follows:

$$[de/dt]_{00} = + \overset{''}{0}.1581330 \qquad [d\Omega/dt]_{00} = - \overset{''}{8}.314194$$
$$[d\chi/dt]_{00} = +12.47677 \qquad [d\pi/dt]_{00} = +12.47244$$
$$[di/dt]_{00} = - 0.2565480 \qquad [dL/dt]_{00} = -18.45083$$

ACTION OF SATURN ON MARS.

E	A	$B \sin \epsilon$	$B \cos \epsilon$	g	h
0°	92.2326164	−11.747198	− 0.835808	39.461155	90.7084011̄
30	93.1075498	−13.263728	+ 6.495252	50.307460	90.7105920
60	94.0543836	−10.902482	+13.616320	33.990063	90.7106532̄
90	94.8268069	− 5.296150	+18.619311	8.020876	90.7086748
120	95.2215463	+ 2.053054	+20.163679	1.205319	90.7070348̄
150	95.1291338	+ 9.175915	+17.835613	24.076867	90.7078833̄
180	94.5669406	+14.163873	+12.258914	57.367408	90.7108082̄
210	93.6819094	+15.680401	+ 4.927852	70.309758	90.7131088
240	92.7148797	+13.319160	− 2.193216	50.728826	90.7125327
270	91.9323586	+ 7.712824	− 7.196207	17.010933	90.7095237̄
300	91.5477178	+ 0.363621	− 8.740573	0.037809	90.7066608̄
330	91.6603255	− 6.759241	− 6.412507	13.064630	90.7063166
Σ_1	560.3380844*	+ 7.250028†	+34.269316‡	182.790580	544.2560900̄
Σ_2	560.3380840	+ 7.250021	+34.269314	182.790524	544.2560991

* $6a^2 + 3a^2e^2 + 6[a'^2 - 2kaa'ee' \cos K] = + 560.3380843.$

† $- 6k'aa' \cos \varphi' \cdot e \sin K' = + 7.250024.$

‡ $6[a'^2e' - kaa'e \cos K] = + 34.269314.$

ACTION OF SATURN ON MARS.

E	l	G	G'	G''	θ		
					°	′	″
0°	1.238258$\overline{1}$	90.7035382	1.5278682	0.2847474	8	6	50.425
30	2.1110005	90.7043316	2,3529760	0.2357148	9	42	47.930
60	3.057773$\overline{2}$	90.7063778	3.1798913	0.1178425	10	59	5.400
90	3.8321748	90.7076570	3.8561238	0.0229312	11	55	59.135
120	4.228554$\overline{3}$	90.7068811	4.2318481	0.0031400	12	28	42.340
150	4.135293$\overline{3}$	90.7048170	4.2015371	0.0631774	12	31	7.643
180	3.570175$\overline{2}$	90.7035495	3.7462608	0.1688274	11	58	46.967
210	2.6828433	90.7043024	2.9540530	0.2624034	10	50	18.088
240	1.7163897	90.7062481	2.0020241	0.2793497	9	6	39.626
270	0.936877$\overline{7}$	90.7074346	1.1081936	0.1692270	6	48	32.560
300	0.555099$\overline{8}$	90.7066561	0.5558543	0.0007499	4	29	34.176
330	0.6680516	90.7047169	0.8409314	0.1712800	6	3	29.499
Σ_1	14.366250$\overline{1}$	544.2332508	15.2437468	0.8546569	57	9	38.934
Σ_2	14.3662411	544.2332595	15.3138149	0.9247338	57	52	14.855

E	$\log K_0$	$\log L_0'$	$\log N_0$	$\log N$	$\log P$	$\log Q$
0°	0.00656257	0.28174175	0.18592136	7.5317335	3.895504$\overline{3}$	5.758669$\overline{4}$
30	0.00942345	0.28554606	0.19019807	7.5468281	3.914863$\overline{9}$	5.778271$\overline{1}$
60	0.01207449	0.28906810	0.19415647	7.5819826	3.9546474	5.817937$\overline{5}$
90	0.01427105	0.29198398	0.19743287	7.6263318	4.0028084	5.8660110
120	0.01561994	0.29377356	0.19944338	7.6674180	4.045881$\overline{0}$	5.9092061
150	0.01572227	0.29390930	0.19959586	7.6949830	4.0730267	5.936646$\overline{1}$
180	0.01438391	0.29213375	0.19760114	7.7028806	4.0781506	5.942049$\overline{9}$
210	0.01175197	0.28863978	0.19367515	7.689588$\overline{2}$	4.060463$\overline{0}$	5.9243808
240	0.00828439	0.28403181	0.18849593	7.658106$\overline{4}$	4.024192$\overline{8}$	5.887629$\overline{6}$
270	0.00461510	0.27914998	0.18300711	7.6156278	3.9778730	5.8401825
300	0.00200564	0.27567456	0.17909843	7.572752$\overline{2}$	3.9331412	5.7942078
330	0.00365091	0.27786617	0.18156337	7.541514$\overline{5}$	3.902482$\overline{3}$	5.764628$\overline{7}$
Σ_1	0.05893094	1.71642353	1.14471671	5.7148732	3.931517$\overline{2}$	5.109700$\overline{1}$
Σ_2	0.05943475	1.71709527	1.14547243	5.7148733	3.931517$\overline{2}$	5,110120$\overline{1}$

ACTION OF SATURN ON MARS.

E	$\log V$	$J_1{}'$	J_2	J_3	F_2	F_3
0°	5.7569714	90.9876364	−0.64905242	+0.2442109	+59.776028	−2.2333582
30	5.7768667	90.9027545	−0.80993896	−1.6448282	+67.492953	−2.2386206
60	5.8172358̄	90.7031717	−0.67748734	−3.0908929	+55.477667	−1.0572468
90	5.8658745	90.5722076	−0.29321392	−3.7064154	+26.949646	+0.0320703
120	5.9091874	90.5928360	+0.18262899	−3.3264146	−10.447038	−0.2313819
150	5.9362703	90.7254544	+0.58278840	−2.0527628	−46.691968	−1.7836963
180	5.9410456	90.8717164	+0.80450796	−0.2268316	−72.073367	−3.2467872
210	5.9228190	90.9409399	+0.82304924	+1.6620728	−79.790278	−3.2597888
240	5.8859649	90.8876171	+0.67261531	+3.1078694	−67.775016	−1.8125358
270	5.8391722	90.7221965	+0.40924575	+3.7232565	−39.246977	−0.2549584
300	5.7942033	90.5707339	+0.07363159	+3.3433908	− 1.850296	+0.0267586
330	5.7636057	90.8202137	−0.30508990	+2.0700081	+34.394639	−1.0495556
Σ_1	5.1046084̄	544.6137115*	+0.40684409	+0.0513320	−36.892022	−8.5545513
Σ_2	5.1046084	544.6837666	+0.40684061	+0.0513310	−36.891985	−8.5545494

E	R_0	$1000000 \times S_0$	$1000 \times W_0$	$1000 \times R^{(n)}$	$100000 \times S^{(n)}$	$1000 \times [R_0 \sin v + (\cos v + \cos E)S_0]$
0°	0.0017097618	+ 9.903284	+0.01219946	0.0000000	+0.7168085	+0.0198066
30	0.0017745243	+ 7.025377	−0.10023833	+0.6334785	+0.5015908	+0.9730086
60	0.0019262900	+ 5.499323	−0.20387007	+1.1484062	+0.3785755	+1.7472855
90	0.0021375714	+ 5.593738	−0.27213044	+1.4028896	+0.3671175	+2.1277319
120	0.0023569201	+ 3.205770	−0.27013289	+1.2799221	+0.2010205	+1.9382857
150	0.0025168546	− 4.917236	−0.17936963	+0.7641820	−0.2986000	+1.1679255
180	0.0025606191	−16.044422	−0.02369074	0.0000000	−0.9631644	+0.0320888
210	0.0024720520	−22.805150	+0.13539869	−0.7505788	−1.3848465	−1.0986739
240	0.0022843230	−19.928906	+0.23709997	−1.2404980	−1.2496588	−1.8606368
270	0.0020608533	− 9.038484	+0.25685358	−1.3525394	−0.5931965	−2.0510270
300	0.0018652798	+ 2.997952	+0.20817962	−1.1120336	+0.2063802	−1.6842325
330	0.0017413378	+ 9.774776	+0.11927113	−0.6216313	+0.6978895	−0.9263634
Σ_1	0.0127031938	−14.366999	−0.04021465	+0.0757967	−0.7100385	+0.1925973
Σ_2	0.0127031934	−14.366879	−0.04021500	+0.0758006	−0.7100452	+0.1926017

* $\Sigma_1(J_1{}' - G'') = 543.7590546$,
$\Sigma_2(J_1{}' - G'') = 543.7590328$.

E	$1000 \times \left[- R_0 \cos v + \left(\frac{r}{a} \sec^2 \varphi + 1 \right) \sin v S_0 \right]$	$1000 \times W_0 \cos u$	$1000 \times W_0 \sin u$	$1000 \times -2\frac{r}{a} R_0$
0°	−1.7097618	+0.00313672	−0.01178931	− 3.1005914
30	−1.4844381	−0.07412689	+0.06747540	− 3.2623831
. 60	−0.8120508	−0.20054993	+0.03664321	− 3.6729182
90	+0.2105546	−0.25530900	−0.09419278	− 4.2751426
120	+1.3414125	−0.17569179	−0.20519295	− 4.9336670
150	+2.2292256	−0.03890712	−0.17509913	− 5.4402943
180	+2.5606191	+0.00609135	−0.02289425	− 5.5988851
210	+2.2161495	−0.09117052	+0.10010366	− 5.3434506
240	+1.3285850	−0.22331936	+0.07965468	− 4.7817011
270	+0.2102889	−0.25329570	−0.04260337	− 4.1217062
300	−0.8010985	−0.15911711	−0.13424046	− 3.5565885
330	−1.4740745	−0.03664087	−0.11350351	− 3.2013711
Σ_1	+1.9077055	−0.74945012	−0.25781908	−25.6443513
Σ_2	+1.9077060	−0.74945010	−0.25781973	−25.6443479

$$\sin \varphi \cdot \tfrac{1}{2} A_1{}^{(s)} + \cos \varphi \cdot B_0{}^{(c)} = + 0.000000000031.$$

Differential Coefficients.

			log coeff.
$[de/dt]_{00}$	$= +$	$22.022051\ m'$	$p\ 1.3428578$
$[d\chi/dt]_{00}$	$= +$	$2338.7360\ \ m'$	$p\ 3.3689812$
$[di/dt]_{00}$	$= -$	$86.444970\ m'$	$n\ 1.936739\overline{8}$
$[d\Omega/dt]_{00}$	$= -$	$920.85894\ m'$	$n\ 2.9641931$
$[d\pi/dt]_{00}$	$= +$	$2338.2557\ \ m'$	$p\ 3.3688920$
$[dL/dt]_{00}$	$= -$	$2935.3283\ \ m'$	$n\ 3.4676567$

Final Values Corresponding to the Above Value of m'.

$$[de/dt]_{00} = +0.0062891406$$
$$[d\chi/dt]_{00} = +0.66790508$$
$$[di/dt]_{00} = -0.024687281$$
$$[d\Omega/dt]_{00} = -0.26298236$$
$$[d\pi/dt]_{00} = +0.66776785$$
$$[dL/dt]_{00} = -0.83828212$$

COMPARISON WITH OTHER RESULTS.

	Leverrier.	Newcomb.	Method of Gauss.
$[de/dt]_{00}$	$+0.00627''$	$+0.00629''$	$+0.0062891''$
$e[d\pi/dt]_{00}$	$+0.06226$	$+0.06226$	$+0.0622814$
$[di/dt]_{00}$	-0.02467	-0.02468	-0.0246873
$\sin i \, [d\Omega/dt]_{00}$	-0.00852	-0.00849	-0.0084927
$[dL/dt]_{00}$	-0.838		-0.8382821

NOTES.

As in other similar cases, the great disagreement of the sums of the functions near the beginning of the computation arises principally from the term a'^2e', but the remarkably close agreement of the final sums shows that the expansion of the perturbing function for this case is very convergent. The greatest effect produced by all terms from the sixth to the eleventh orders inclusive here occurs with $[de/dt]_{00}$ and amounts to but 1/100000th of the value of this coefficient.

Dr. Samuel G. Barton has effected this computation from the same elements as are here employed, making use of the formulas developed by Dr. Arndt[30]. (*A Thesis presented to the Faculty of the Graduate School of the University of Pennsylvania, 1906*). The results obtained by him are as follows:

$$[de/dt]_{00} = +0.0062897''$$
$$e[d\pi/dt]_{00} = +0.0622817$$
$$[di/dt]_{00} = -0.0246873$$
$$\sin i \, [d\Omega/dt]_{00} = -0.0084927$$
$$[dL/dt]_{00} = -0.8382857$$

The agreement is thus practically exact.

It is the conclusion of Dr. Barton that in spite of the greater symmetry of the equations employed in the method of Arndt, computations effected by them are somewhat less accurate than when the methods of Hill are employed. His residual arising from the equation derived from the constancy of the major axis is eight times greater than that here obtained. (See the notes to the computation of the action of Jupiter on Mars, where it is shown that Dr. Turner came to the same conclusion.)

ACTION OF URANUS ON MARS.

E	A	B cos ε	B sin ε	g	h
0°	367.811028$\overline{3}$	− 8.331256	+ 6.828502	37.81310	367.4959$\overline{8}$
45	368.2057075	− 5.369694	−15.243795	188.44191	367.49649
90	370.169885$\overline{6}$	+12.404411	−28.797937	672.53375	367.5004$\overline{4}$
135	372.5672534	+34.579222	−25.894092	543.74175	367.5003$\overline{0}$
180	373.9791848	+48.165033	− 8.233286	54.97149	367.49639
225	373.5643097	+45.203469	+13.839014	155.31082	367.49635
270	371.5799361	+27.429370	+27.393154	608.52085	367.49979
315	369.2027636	+ 5.254555	+24.489302	486.34483	367.49949
Σ₁	1483.5400347*	+79.667558†	− 2.809567‡	1373.83919	1469.99259
Σ₂	1483.5400342	+79.667552	− 2.809571	1373.83931	1469.9926$\overline{3}$

* $4a^2 + 2a^2e^2 + 4[a'^2 − 2kaa'ee' \cos K] = 1483.5400348.$

† $4[a'^2e' − kaa'e \cos K] = + 79.667564.$

‡ $− 4k'aa' \cos \varphi' \cdot e \sin K' = − 2.809567.$

E	l	G	G'	G''	θ
0°	− 0.49589	367.4957$\overline{0}$	0.157753$\overline{6}$	0.653148	2° 41′ 22.746″
45	− 0.10173	367.4951$\overline{0}$	0.667671	0.768004	3 3$\overline{4}$ 47.145
90	+ 1.8585$\overline{1}$	367.49543	2.574380	0.710869	5 2$\overline{5}$ 12.441
135	+ 4.25601	367.49622	4.582935	0.322847	6 37 54.392
180	+ 5.67185	367.4959$\overline{8}$	5.698515	0.026250	7 10 10.482
225	+ 5.2570$\overline{2}$	367.4951$\overline{9}$	5.33736$\overline{2}$	0.079182$\overline{}$	6 58 20.716
270	+ 3.26920	367.4952$\overline{5}$	3.718991	0.445245	6 6 25.010
315	+ 0.89233	367.49588	1.682505	0.786567$\overline{}$	4 41 47.833
Σ₁	+10.3036$\overline{7}$	1469.9823$\overline{5}$	12.149421	1.835512	21 23 10.679
Σ₂	+10.3036$\overline{3}$	1469.98238	12.27047$\overline{3}$	1.956599	21 52 50.086

ACTION OF URANUS ON MARS.

E	log K_0	log L_0'	log N_0	log N	log P	log Q
0°	0.00071814	0.27395868	0.17716832	6.6153321	1.7572439	4.226477$\overline{0}$
45	0.00127262	0.27469773	0.17799967	6.6414665	1.7838479	4.253308$\overline{1}$
90	0.00292086	0.27689384	0.18046982	6.702475$\overline{3}$	1.8471866	4.3168538
135	0.00437720	0.27883326	0.18265093	6.7600912	1.907656$\overline{1}$	4.3771079
180	0.00511859	0.27982020	0.18376077	6.7833377	1.9325907	4.4018149
225	0.00484000	0.27944936	0.18334377	6.7609874	1.909746$\overline{3}$	4.378986$\overline{0}$
270	0.00371007	0.27794494	0.18165196	6.7037350	1.850124$\overline{8}$	4.319609$\overline{4}$
315	0.00219200	0.27592287	0.17937772	6.642351$\overline{6}$	1.7859125	4.2555483
Σ₁	0.01246766	1.10861766	0.72305087	6.804880$\overline{1}$	7.387146$\overline{0}$	7.2647550
Σ₂	0.01268182	1.10890322	0.72337209	6.8048967	7.3871627	7.2649502

ACTION OF URANUS ON MARS.

E	$\log V$	$J_1{}'$	J_2	J_3	F_2
0°	4.2255130	368.03012$\bar{6}$	+0.3910957	+5.9910041	−117.85624
45	4.2521750	368.25794$\bar{3}$	−0.6874254	+1.3122864	+263.09961
90	4.3158056	368.150829	−1.4287110	−4.4833753	+497.03672
135	4.3766319	367.663776	−1.2456453	−8.0008361	+446.91789
180	4.4017762	367.40322$\bar{8}$	−0.3170169	−7.1797383	+142.10202
225	4.3788692	367.55909$\bar{0}$	+0.6967658	−2.5011942	−238.85387
270	4.318952$\bar{9}$	367.91123$\bar{0}$	+1.2249811	+3.2942931	−472.79092
315	4.2543883	368.137688	+1.1030033	+6.8119278	−422.67209
Σ_1	7.262047$\bar{7}$	1471.49541$\bar{2}$*	−0.1296511	−2.3778169	+ 48.49158
Σ_2	7.2620644	1471.61849$\bar{6}$	−0.1333016	−2.3778161	+ 48.49154

E	F_3	$1000 \times R_0$	$1000000 \times S_0$	$1000000 \times W_0$	$1000 \times R^{(n)}$	$1000000 \times S^{(n)}$
0°	+ 1.1362797	0.2059026	−0.0165525	+10.07610	0.0000000	−0.0119808
45	− 5.0397791	0.2189646	+0.3708684	+ 2.31469	+0.1087907	+0.2605870
90	− 8.3838519	0.2523915	+0.5397194	− 9.33604	+0.1656447	+0.3542182
135	− 1.8404722	0.2886690	+0.6481209	−19.05928	+0.1256756	+0.3990450
180	+ 1.6518880	0.3050752	+0.4171470	−18.09445	0.0000000	+0.2504179
225	− 5.0490913	0.2898966	−0.2732646	− 6.02536	−0.1262101	−0.1682478
270	− 8.9126276	0.2535451	−0.7948804	+ 6.80306	−0.1664018	−0.5216806
315	− 2.5789640	0.2193839	−0.6003932	+12.22078	−0.1089990	−0.4218605
Σ_1	−14.5083118	1.0169144	+0.1454335	−10.55133	−0.0007571	+0.0709747
Σ_2	−14.5083066	1.0169141	+0.1453315	−10.54917	−0.0007428	+0.0695237

E	$1000 \times [R_0 \sin v + (\cos v + \cos E)S_0]$	$1000 \times \left[-R_0 \cos v + \left(\dfrac{r}{a} \sec^2 \varphi + 1\right) \sin v S_0\right]$	$1000000 \times W_0 \cos u$	$1000000 \times W_0 \sin u$	$1000 \times -2\dfrac{r}{a} R_0$
0°	−0.00003310	−0.20590256	+2.590763	− 9.737343	−0.37339681
45	+0.16554695	−0.14335624	+2.077126	− 1.021440	−0.40904755
90	+0.25124093	+0.02461951	−8.758941	− 3.231492	−0.50478291
135	+0.18971131	+0.21763709	−8.485224	−17.066255	−0.61541380
180	−0.00083429	+0.30507521	+4.652436	−17.486113	−0.66705785
225	−0.19106859	+0.21804505	+5.009009	− 3.348852	−0.61803079
270	−0.25236576	+0.02523742	−6.708828	− 1.128399	−0.50709023
315	−0.16617611	−0.14329578	−6.836536	−10.129621	−0.40983085
Σ_1	−0.00199222	+0.14902958	−8.224570	−31.583347	−2.05232780
Σ_2	−0.00198644	+0.14903012	−8.235625	−31.566168	−2.05232299

$\sin \varphi \cdot \tfrac{1}{2}A_1{}^{(s)} + \cos \varphi \cdot B_0{}^{(c)} = -0.0000000000013.$

* $\Sigma_1(J_1{}' - G'') = 1469.65990\bar{0}.$
$\Sigma_2(J_1{}' - G'') = 1469.66189\bar{7}.$

DIFFERENTIAL COEFFICIENTS.

			log coeff.
$[de/dt]_{00}$	$= -$	$0.34119354\ m'$	$n\ 9.5330008$
$[d\chi/dt]_{00}$	$= +274.05283$	m'	$p\ 2.4378343$
$[di/dt]_{00}$	$= -$	$1.4239452\ m'$	$n\ 0.1534933$
$[d\Omega/dt]_{00}$	$= -169.16430$	m'	$n\ 2.2283087$
$[d\pi/dt]_{00}$	$= +273.96460$	m'	$p\ 2.4376945$
$[dL/dt]_{00}$	$= -352.43262$	m'	$n\ 2.5470761$

FINAL VALUES CORRESPONDING TO THE ABOVE VALUE OF m'.

$$[de/dt]_{00} = -0.000014964631$$
$$[d\chi/dt]_{00} = +0.012019862$$
$$[di/dt]_{00} = -0.000062453743$$
$$[d\Omega/dt]_{00} = -0.0074194879$$
$$[d\pi/dt]_{00} = +0.012015994$$
$$[dL/dt]_{00} = -0.015457573$$

COMPARISON WITH OTHER RESULTS.

	Leverrier.	Newcomb.	Method of Gauss.
$[de/dt]_{00}$	-0.00001	-0.00001	-0.000014964631
$e[d\pi/dt]_{00}$	$+0.00112$	$+0.00112$	$+0.0011207080$
$[di/dt]_{00}$	-0.00007	-0.00006	-0.000062453743
$\sin i\ [d\Omega/dt]_{00}$	-0.00023	-0.00024	-0.00023960370
$[dL/dt]_{00}$	-0.015		-0.015457573

NOTES.

The greatest error produced in this case by a division into only four parts occurs with the coefficient $[d\chi/dt]_{00}$ and amounts to but $0''.0000001$. It is evident that, notwithstanding the disagreement of the sums of the functions in the first part of the computation, a division into eight parts is fully sufficient.

ACTION OF NEPTUNE ON MARS.

E	A	$B \cos \epsilon$	$B \sin \epsilon$	g	h
0_o	906.38891911	$+21.92636$	-39.01745	99.36880	904.17365
45	906.94215394	$+47.61371$	-15.38170	15.44333	904.17339
90	907.26271705	$+49.05172$	$+19.61543$	25.11471	904.17633
135	907.17710690	$+25.39796$	$+45.47311$	134.97137	904.17696
180	906.72119273	$- 9.49146$	$+47.04426$	144.45927	904.17395
225	906.14776209	-35.17881	$+23.40851$	35.76678	904.17317
270	905.80700342	-36.61681	-11.58863	8.76590	904.17576
315	905.01280912	-12.96307	-37.44631	91.52721	904.17627
Σ_1	3626.17983231*	$+24.86981$†	$+16.05361$‡	277.70868	3616.69969
Σ_2	3626.17983205	$+24.86979$	$+16.05361$	277.70869	3616.69979

E	l	G	G'	G''	θ		
					°	′	″
$0°$	2.15000	904.17353	2.20007	0.049953	2	51	33.458
45	2.70349	904.17337	2.70981	0.006303	3	8	30.703
90	3.02112	904.17630	3.03031	0.009166	3	19	25.738
135	2.93488	904.1768$\bar{0}$	2.98505	0.050008	3	19	16.746
180	2.48197	904.17377	2.54493	0.062779	3	4	42.112
225	1.90932	904.1731$\bar{3}$	1.92986	0.020498	2	39	43.146
270	1.56597	904.17575	1.57215	0.006167	2	23	40.258
315	1.67127	904.17616	1.7299$\bar{0}$	0.058516	2	32	56.175
Σ_1	9.21906	3616.69935	9.34746	0.128065	11	39	21.566
Σ_2	9.21896	3616.69945	9.3546$\bar{2}$	0.135325	11	40	26.770

ACTION OF NEPTUNE ON MARS.

E	$\log K_0$	$\log L_0'$	$\log N_0$	$\log N$	$\log P$	$\log Q$
$0°$	0.00081164	0.27408332	0.17730851	6.030047$\bar{1}$	0.391578$\bar{8}$	3.251079$\bar{8}$
45	0.00098011	0.27430787	0.17756111	6.0560291	0.417827$\bar{5}$	3.277335$\bar{5}$
90	0.00109700	0.27446368	0.17773639	6.115402$\bar{1}$	0.477350$\bar{8}$	3.336881$\bar{0}$
135	0.00109536	0.27446148	0.17773391	6.1708446	0.5327514	3.392301$\bar{2}$
180	0.00094087	0.27425556	0.17750228	6.1926623	0.554353$\bar{8}$	3.413882$\bar{6}$
225	0.00070344	0.27393908	0.17714626	6.1704766	0.5318928	3.3913614
270	0.00056913	0.27376004	0.17694484	6.114876$\bar{8}$	0.4761252	3.3355658
315	0.00064494	0.27386109	0.17705853	6.0556543	0.4169531	3.2764317
Σ_1	0.00341864	1.09656260	0.70949202	4.452988$\bar{2}$	1.899408$\bar{5}$	3.337409$\bar{1}$
Σ_2	0.00342385	1.09656952	0.70949981	4.4530046	1.899424$\bar{8}$	3.3374297

* $4a^2 + 2a^2e^2 + 4[a'^2 - 2kaa'ee' \cos K] = 3626.17983218$.

† $4[a'^2e' - kaa'e \cos K] = + 24.869793$.

‡ $- 4k'aa' \cos \varphi' \cdot e \sin K' = + 16.05361$.

Action of Neptune on Mars.

E	$\log V$	J_1'	J_2	J_3	F_2
0°	3.2510498̄	903.898094	+0.3009875	+15.528734	+299.48600
45	3.2773317	903.965295	−0.6661748	−12.963466	+118.06523
90	3.3368755̄	902.840060	−0.3701497	−34.790528	−150.56206
135	3.3922712̄	902.874950	+0.9173304	−37.166389	−349.03766
180	3.4138449̄	903.910920	+1.0320306	−18.699362	−361.09725
225	3.3913491	904.099942	−0.1708692	+ 9.792741	−179.67656
270	3.3355621	903.071970	−0.8044972	+31.619710	+ 88.95077
315	3.2763966	902.764465	+0.0451253	+33.995664	+287.42636
Σ_1	3.3373322̄	3613.721044*	+0.1583712	− 6.341446	−123.22254
Σ_2	3.3373486̄	3613.704652	+0.1254117	− 6.341450	−123.22263

E	F_3	$1000 \times R_0$	$1000000 \times S_0$	$1000000 \times W_0$	$1000 \times R^{(n)}$	$1000000 \times S^{(n)}$
0°	−10.033551	0.05357158	+0.12743633	+2.7656535	0.00000000	+0.09223957
45	− 4.253985	0.05690845	−0.09526013	−2.4561209	+0.02827449	−0.06693360
90	+ 2.323772	0.06500834	−0.12559190	−7.5560767	+0.04266503	−0.08242607
135	− 5.662284	0.07385617	+0.10733754	−9.1730340	+0.03215420	+0.06608723
180	−14.586465	0.07791030	+0.13821863	−4.8544449	0.00000000	+0.08297417
225	− 7.712269	0.07407227	−0.10322243	+2.4086872	−0.03224828	−0.06355357
270	+ 1.985936	0.06494182	−0.14759132	+6.8479047	−0.04262137	−0.09686429
315	− 2.681773	0.05659771	+0.08359968	+6.4235154	−0.02812010	+0.05874051
Σ_1	−20.310308	0.26143204	−0.00752826	−2.7969634	+0.00004366	−0.00407662
Σ_2	−20.310311	0.26143460	−0.00754534	−2.7969523	+0.00006031	−0.00565943

E	$1000 \times [R_0 \sin v + (\cos v + \cos E)S_0]$	$1000 \times \left[-R_0 \cos v + \left(\frac{r}{a} \sec^2 \varphi + 1 \right) \sin v \cdot S_0 \right]$	$1000000 \times W_0 \cos u$	$1000000 \times W_0 \sin u$	$1000 \times -2\frac{r}{a} R_0$
0°	+0.000254873	−0.053571580	+ 0.7111033	− 2.6726716	−0.09715013
45	+0.042763861	−0.037538560	−. 2.2040406	+ 1.0838516	−0.10631064
90	+0.064736687	+0.005812013	− 7.0890057	− 2.6153925	−0.13001668
135	+0.048623034	+0.055602449	− 4.0838509	− 8.2138132	−0.15745404
180	−0.000276437	+0.077910304	+ 1.2481721	− 4.6912376	−0.17035369
225	−0.048771753	+0.055759058	− 2.0023922	+ 1.3387309	−0.15791475
270	−0.064644985	+0.006352181	− 6.7530485	− 1.1358370	−0.12988365
315	−0.042545549	−0.037317274	− 3.5934362	− 5.3243555	−0.10573015
Σ_1	+0.000070138	+0.036502918	−11.8827788	−11.1151387	−0.52740415
Σ_2	+0.000069593	+0.036505673	−11.8837299	−11.1155862	−0.52740958

$\sin \varphi \cdot \frac{1}{2}A_1^{(s)} + \cos \varphi \cdot B_0^{(c)} = + 0.0000000000012.$

* $\Sigma_1(J_1' - G'') = 3613.592979.$

$\Sigma_2(J_1' - G'') = 3613.569327.$

Differential Coefficients.

log coeff.

$$[de/dt]_{00} = + 0\overset{''}{.}011982 \; m' \quad p\; 8.078535\bar{0}$$
$$[d\chi/dt]_{00} = +67.128215 \; m' \quad p\; 1.8269051$$
$$[di/dt]_{00} = - 2.0560028 \; m' \quad n\; 0.3130237$$
$$[d\Omega/dt]_{00} = -59.551438 \; m' \quad n\; 1.7748923$$
$$[d\pi/dt]_{00} = +67.097154 \; m' \quad p\; 1.8267041$$
$$[dL/dt]_{00} = -90.590942 \; m' \quad n\; 1.9570848$$

Final Values Corresponding to the Above Value of m'.

$$[de/dt]_{00} = +0\overset{''}{.}00000060823$$
$$[d\chi/dt]_{00} = +0.0034075236$$
$$[di/dt]_{00} = -0.00010436562$$
$$[d\Omega/dt]_{00} = -0.0030229161$$
$$[d\pi/dt]_{00} = +0.0034059472$$
$$[dL/dt]_{00} = -0.0045985255$$

Comparison with Other Results.

	Leverrier.	Newcomb.	Method of Gauss.
$[de/dt]_{00}$	$+0\overset{''}{.}00000$	$+0\overset{''}{.}00000$	$+0\overset{''}{.}00000060823$
$e[d\pi/dt]_{00}$	$+0.00032$	$+0.00032$	$+0.00031766599$
$[di/dt]_{00}$	-0.00011	-0.00011	-0.00010436562
$\sin i \,[d\Omega/dt]_{00}$	-0.00009	-0.00010	-0.000097621545
$[dL/dt]_{00}$	-0.004		-0.0045985255

Notes.

The agreement of the sums of the functions is much more exact throughout than in the preceding computation because e' is here so much smaller. The greatest effect produced by all terms from the fourth to the seventh orders inclusive is but $0''.000001$, and it is evident that the terms of the eighth and higher orders are wholly inappreciable.

11. THE FINAL VALUES OF THE PERTURBATIONS.

Combining the results of the preceding pages, we now obtain the values of the perturbations stated in the following tables. For comparison with these, the results obtained by LEVERRIER[7] and NEWCOMB[15] are added, all of the results being reduced to the values of the masses here adopted and stated in Article 6.

SECULAR PERTURBATIONS OF MERCURY.

(Epoch 1850.0, G. M. T.)

Action of—	$\left[\dfrac{de}{dt}\right]_{00}$	$\left[\dfrac{d\chi}{dt}\right]_{00}$	$\left[\dfrac{di}{dt}\right]_{00}$
Venus...............	$+0.027739414$	$+2.7763615$	-0.14811133
Earth...............	$+0.011476557$	$+0.91448833$	-0.014040890
Mars...............	-0.000607428	$+0.02486334$	-0.000301945
Jupiter.............	$+0.00319413$	$+1.5400720$	-0.049056191
Saturn..............	$+0.000531095$	$+0.07312263$	-0.004212776
Uranus.............	$+0.000009638$	$+0.00142135$	-0.000024450
Neptune............	$+0.000003320$	$+0.00041901$	-0.000020027
	-0.04234673	$+5.3307482$	-0.21576761

Action of—	$\left[\dfrac{d\Omega}{dt}\right]_{00}$	$\left[\dfrac{d\pi}{dt}\right]_{00}$	$\left[\dfrac{dL}{dt}\right]_{00}$
Venus...............	-1.9420214	$+2.7618772$	-3.2505323
Earth...............	-1.0037245	$+0.90700208$	-1.1935233
Mars...............	-0.01926435	$+0.02471966$	-0.03293324
Jupiter.............	-1.4795642	$+1.5290366$	-2.2066350
Saturn..............	-0.06979662	$+0.07260205$	-0.10657405
Uranus.............	-0.00134987	$+0.00141128$	-0.00201139
Neptune............	-0.00044314	$+0.00041570$	-0.00060031
	-4.5161641	$+5.2970646$	-6.7928096

COMPARISON WITH OTHER RESULTS.

	Leverrier.	Newcomb.	Method of Gauss.
$[de/dt]_{00}$	$+0.04246$	$+0.04234$	$+0.0423467$
$e[d\pi/dt]_{00}$	$+1.08946$	$+1.09601$	$+1.0891018$†
$[di/dt]_{00}$	-0.21586	-0.21570	-0.2157676
$\sin i\,[d\Omega/dt]_{00}$	-0.55017	-0.55041	-0.5505495
$[dL/dt]_{00}$	$-6.8190*$		-6.79281

* Exclusive of the action of Uranus and Neptune.

† This unexpectedly large difference is a gradual accumulation from all of the computations Thus, the residuals, Newcomb–Gauss, are, in the several cases: 0″.00300, 0″.00151, 0″.00003, 0″.00227, and 0″.00010, the sum of which is the difference as found above.

SECULAR PERTURBATIONS OF VENUS.

(Epoch 1850.0, G. M. T.)

Action of—	$\left[\dfrac{de}{dt}\right]_{00}$	$\left[\dfrac{d\chi}{dt}\right]_{00}$	$\left[\dfrac{di}{dt}\right]_{00}$
Mercury	$-0.\overset{\prime\prime}{0}13012279$	$-1.\overset{\prime\prime}{1}893992$	$+0.\overset{\prime\prime}{0}094965089$
Earth	-0.04898290	-5.6289701	$+0.000044940$
Mars	-0.001963988	$+0.74594759$	$+0.0013204280$
Jupiter	-0.031162921	$+6.5654682$	-0.038659982
Saturn	-0.000675363	$+0.07935156$	-0.0052327048
Uranus	$+0.000005263$	$+0.00278176$	$+0.0000018240$
Neptune	-0.000000278	$+0.00110440$	-0.0000283988
	-0.09579247	$+0.5762842$	-0.033057385

Action of—	$\left[\dfrac{d\Omega}{dt}\right]_{00}$	$\left[\dfrac{d\pi}{dt}\right]_{00}$	$\left[\dfrac{dL}{dt}\right]_{00}$
Mercury	$+\ 0.\overset{\prime\prime}{0}897732$	$-1.\overset{\prime\prime}{1}892420$	$+\ 0.\overset{\prime\prime}{7}4542525$
Earth	$-\ 7.293993$	-5.6417558	$-\ 5.4005288$
Mars	$-\ 0.0473504$	$+0.74586465$	$-\ 0.09940123$
Jupiter	$-\ 2.7242270$	$+6.5606924$	$-\ 5.5347410$
Saturn	$-\ 0.0824657$	$+0.07920700$	$-\ 0.26491624$
Uranus	$-\ 0.0028813$	$+0.00277671$	$-\ 0.00496096$
Neptune	$-\ 0.0007780$	$+0.00110304$	$-\ 0.00148569$
	-10.061922	$+0.5586460$	-10.5606087

COMPARISON WITH OTHER RESULTS.

	Leverrier.	Newcomb.	Method of Gauss.
$[de/dt]_{00}$	$-\ 0.\overset{\prime\prime}{0}9558$	$-0.\overset{\prime\prime}{0}9576$	$-\ 0.0957925$
$e[d\pi/dt]_{00}$	$+\ 0.00366$	$+0.00392$	$+\ 0.0038229$
$[di/dt]_{00}$	$-\ 0.03318$	-0.03306	$-\ 0.0330574$
$\sin i\ [d\Omega/dt]_{00}$	$-\ 0.59530$	-0.59551	$-\ 0.5955192$
$[dL/dt]_{00}$	-10.549		-10.5606087

SECULAR PERTURBATIONS OF THE EARTH.

(Epoch 1850.0, G. M. T.)

Action of—	$\left[\dfrac{de}{dt}\right]_{00}$	$\left[\dfrac{d\pi}{dt}\right]_{00} = \left[\dfrac{d\chi}{dt}\right]_{00}$	$\left[\dfrac{dp}{dt}\right]_{00}$
Mercury	$-0.\overset{\prime\prime}{0}011613570$	$-\ 0.\overset{\prime\prime}{1}0999815$	$+0.\overset{\prime\prime}{0}025085775$
Venus	$+0.013483339$	$+\ 3.4537341$	$+0.074457966$
Mars	-0.015723904	$+\ 0.97519611$	$+0.0063443986$
Jupiter	-0.081841849	$+\ 6.9652565$	-0.025114405
Saturn	-0.0004330571	$+\ 0.18725991$	-0.0054235259
Uranus	$+0.0000172788$	$+\ 0.00566366$	$+0.0000236793$
Neptune	-0.0000006006	$+\ 0.00179708$	-0.0000364953
	-0.085660150	$+11.4789092$	$+0.052760195$

Action of—	$\left[\dfrac{dq}{dt}\right]_{00}$	$\left[\dfrac{dL}{dt}\right]_{00}$
Mercury....................	-0.002098681	$+\ 0.3930935$
Venus......................	-0.28462399	$+11.232473$
Mars.......................	-0.007195311	$-\ 0.2342424$
Jupiter....................	-0.16046446	$-\ 9.1916336$
Saturn.....................	-0.013188086	$-\ 0.4325140$
Uranus....................	-0.0000784873	$-\ 0.0080930$
Neptune...................	-0.0000432488	$-\ 0.0024199$
	-0.46769226	$+\ 1.756664$

COMPARISON WITH OTHER RESULTS.

	Leverrier.	Newcomb.	Method of Gauss.
$[de/dt]_{00}$	-0.08569	-0.08563	-0.085660
$e[d\pi/dt]_{00}$	$+0.19254$	$+0.19248$	$+0.192514$
$[dp/dt]_{00}$	$+0.05290$	$+0.05276$	$+0.052760$
$[dq/dt]_{00}$	-0.46754	-0.46768	-0.467692
$[dL/dt]_{00}$	$+1.7570*$		$+1.756664$

The values of $[dp/dt]_{00}$ and $[dq/dt]_{00}$ obtained by HILL in the "*New Theory*" are given below. These were regarded as provisional results only, and were derived from the numerical values of the coefficients in the expansion of the perturbing function stated by LEVERRIER in the *Annales*, Vol. II.

It may also be of interest to add the results obtained by the first application ever made of the method of GAUSS. This was a computation by NICOLAI of the secular perturbations of the Earth, the final values only being published, in BODE's *Berliner Jahrbuch*, 1820, pages 224–226 (Aug. 30, 1817). These results are here reduced to the values of the masses stated in Article 6.

	Hill.	Nicolai.
$[de/dt]_{00}$		-0.08606
$e[d\pi/dt]_{00}$		$+0.19283$
$[dp/dt]_{00}$	$+0.0527225$	$+0.05182$
$[dq/dt]_{00}$	-0.4676079	-0.46738

* Exclusive of the action of Neptune. If the value of this found above is included, we have $[dL/dt]_{00} = 1''.7546$; —a less exact agreement.

SECULAR PERTURBATIONS OF MARS.

(Epoch 1850.0, G. M. T.)

Action of—	$\left[\dfrac{de}{dt}\right]_{00}$	$\left[\dfrac{d\chi}{dt}\right]_{00}$	$\left[\dfrac{di}{dt}\right]_{00}$
Mercury...............	+0.000335670	+ 0.0061841	+0.000074482
Venus.................	+0.000795405	+ 0.4947286	−0.012829757
Earth.................	+0.021481158	+ 2.2915614	+0.000319911
Jupiter...............	+0.15813453	+12.476799	−0.25654077
Saturn................	+0.006289141	+ 0.6679051	−0.024687281
Uranus................	−0.000014965	+ 0.0120199	−0.000062454
Neptune..............	+0.000000608	+ 0.0034075	−0.000104366
	+0.18702155	+15.952606	−0.29383023

Action of—	$\left[\dfrac{d\Omega}{dt}\right]_{00}$	$\left[\dfrac{d\pi}{dt}\right]_{00}$	$\left[\dfrac{dL}{dt}\right]_{00}$
Mercury...............	+ 0.01479483	+ 0.0061918	+ 0.1940178
Venus.................	+ 0.30877426	+ 0.4948896	+ 4.1204933
Earth.................	− 2.2862242	+ 2.2903688	+ 6.6520970
Jupiter...............	− 8.3142000	+12.472464	−18.450874
Saturn................	− 0.26298236	+ 0.6677678	− 0.8382821
Uranus....	− 0.00741949	+ 0.0120160	− 0.0154576
Neptune..............	− 0.00302292	+ 0.0034059	− 0.0045985
	−10.5502799	+15.947104	− 8.342604

COMPARISON WITH OTHER RESULTS.

	Leverrier.	Newcomb.	Method of Gauss.
$[de/dt]_{00}$	+0.18703	+0.18706	+0.187022
$e[d\pi/dt]_{00}$	+1.48645	+1.48787	+1.487355
$[di/dt]_{00}$	−0.29375	−0.29385	−0.293830
$\sin i\,[d\Omega/dt]_{00}$	−0.34099	−0.34066	−0.340709
$[dL/dt]_{00}$	−8.358*		−8.34260

12. COMPARISON WITH THE RESULTS OF OBSERVATION.

From a discussion of all the available observations of the planets and of the Sun, NEWCOMB has derived the most probable values of the preceding coefficients based upon observations alone. These will be found summarized in a convenient form on pages 107 and 108 of *The Elements of the Four Inner Planets and the Fundamental Constants of Astronomy* (Supplement to the American Ephemeris and Nautical Almanac, 1897).

* The value of $[dL/dt]_{00}$ arising from the action of Mercury was not stated by Leverrier. The value as found above has been added to his series of values in order to obtain this sum.

In order to compare the values here obtained with those given by NEWCOMB it is necessary to notice that the values of i and Ω stated by him are measured from the movable equator and equinox and that it is therefore necessary to free the values of $[di/dt]_{00}$ and $[d\Omega/dt]_{00}$ here given from the changes caused by the motion of the ecliptic itself. For this purpose we first compute ρ and L from the equations,

$$\rho \sin L = \left[\frac{dp}{dt}\right]_{00} \quad \text{and} \quad \rho \cos L = \left[\frac{dq}{dt}\right]_{00},$$

the secular variations being those which belong to the Earth's orbit, and then add the quantities $-\rho \cos (L - \Omega)$ to the several determinations of $[di/dt]_{00}$ and $-\rho \times \cos i \sin (L - \Omega)$ to those above given for $\sin i \, [d\Omega/dt]_{00}$. In this manner the values stated in the following tabulation are obtained.

In a similar way it might appear necessary to add the correction,

$$e \tan \tfrac{1}{2} i \left(\sin i \left[\frac{d\Omega}{dt}\right]_{00} + \rho \sin (L - \Omega) \right)$$

to the values obtained for $e \, [d\pi/dt]_{00}$, the first part arising from the change due to the plane of the orbit and the second from that produced by the motion of the ecliptic. And in the case of the Earth's perihelion, there is a secular motion due to the lack of sphericity of the Earth-moon system which is expressed by the equation,

$$e\left[\frac{d\pi}{dt}\right]_{00} = \tfrac{3}{2}en \cdot \frac{mm'}{(m+m')^2} \cdot \left(\frac{a'}{a}\right)^2,$$

the accented letters applying to the moon (*Annales de l'Observatoire de Paris*, Vol. IV, pages 42–46). Employing the values of a' and m' given in the *Astronomical Papers of the American Ephemeris*, Vol. IV, page 11, this correction is found to be $+0''.0157884$. But these last two corrections need not here be applied because the values of the variations obtained by NEWCOMB from observation have already been freed from their effects.

MERCURY.

	Newcomb.	Method of Gauss.	Observ.	δ_1	δ_2	ϵ
	''	''	''	''	''	''
$[de/dt]_{00}$	$+0.0423$	$+0.0423$	$+0.0336$	-0.0087	-0.0087	±0.0050
$e \, [d\pi/dt]_{00}$	$+1.0960$	$+1.0891$	$+1.1824$	$+0.0864$	$+0.0933$	±0.0040
$[di/dt]_{00}$	$+0.0676$	$+0.0674$	$+0.0714$	$+0.0038$	$+0.0040$	±0.0080
$\sin i \, [d\Omega/dt]_{00}$	-0.9250	-0.9234	-0.9189	$+0.0061$	$+0.0045$	±0.0045

VENUS.

	Newcomb.	Method of Gauss.	Observ.	δ_1	δ_2	ϵ
	$''$	$''$	$''$	$''$	$''$	$''$
$[de/dt]_{00}$	-0.0958	-0.0958	-0.0946	$+0.0012$	$+0.0012$	± 0.0020
$e\ [d\pi/dt]_{00}$	$+0.0039$	$+0.0038$	$+0.0029$	-0.0010	-0.0009	± 0.0020
$[di/dt]_{00}$	$+0.0034$	$+0.0034$	$+0.0029$	-0.0005	-0.0005	± 0.0030
$\sin i\ [d\Omega/dt]_{00}$	-1.0600	-1.0603	-1.0540	$+0.0060$	$+0.0063$	± 0.0012

EARTH.

	Newcomb.	Method of Gauss.	Observ.	δ_1	δ_2	ϵ
	$''$	$''$	$''$	$''$	$''$	$''$
$[de/dt]_{00}$	-0.0856	-0.0857	-0.0855	$+0.0001$	$+0.0002$	± 0.0009
$e\ [d\pi/dt]_{00}$	$+0.1925$	$+0.1925$	$+0.1948$	$+0.0023$	$+0.0023$	± 0.0012
$[d\epsilon/dt]_{00}$	-0.4677	-0.4677	-0.4711	-0.0034	-0.0034	± 0.0023

MARS.

	Newcomb.	Method of Gauss.	Observ.	δ_1	δ_2	ϵ
	$''$	$''$	$''$	$''$	$''$	$''$
$[de/dt]_{00}$	$+0.1871$	$+0.1870$	$+0.1900$	$+0.0029$	$+0.0030$	± 0.0027
$e\ [d\pi/dt]_{00}$	$+1.4879$	$+1.4874$	$+1.4955$	$+0.0076$	$+0.0081$	± 0.0035
$[di/dt]_{00}$	-0.0225	-0.0229	-0.0226	-0.0001	$+0.0003$	± 0.0020
$\sin i\ [d\Omega/dt]_{00}$	-0.7263	-0.7251	-0.7260	$+0.0003$	-0.0009	± 0.0020

In the above tabulation the column headed δ_1 expresses the residuals from the computation of NEWCOMB and that headed δ_2 states the residuals from the results here obtained. It will be noticed that the differences are very minute throughout, the only appreciable improvement arising from the more accurate computation occurring in the case of the node of Mercury, where the residual is reduced by its fourth part.

The last column contains the mean errors of the observational results. If we multiply these by 0.6745 to reduce them to probable errors, we observe that in seven cases the residuals are less than the probable errors; in five cases they vary from one to three times as great but that in each of these cases where the divergence is greatest a slight change in the value of the masses will correct the disagreement, and that in the remaining three cases the difference is very much greater than can be ascribed to errors either in the adopted masses, the computation, or to errors in the observations themselves. These three cases are:

1. The motion of the perihelion of Mercury.
2. The motion in the node of Venus.
3. The motion of the perihelion of Mars.

The first of these is the well-known discordance. The second is well established, the discordance between observation and theory being nearly eight times the probable error, nor can the uncertainty remaining in the values of the masses account for more than a small part of the discrepancy. NEWCOMB estimates the mean error of the computed value arising from this uncertainty as not more than $\pm 0''.0012$, so that with this included the residual is nearly six times the probable error. The third discordance is the least of the three, but as the masses of Jupiter and Saturn, the principal disturbing planets for this case, are accurately known, the uncertainty of the computed results is almost negligible. NEWCOMB estimates the mean error of the result of computation arising from the uncertainties in the masses of all the planets as here but $\pm 0''.0004$, so that the residual remains between three and four times as large as the probable error.

13. COMPARISON WITH SEELIGER'S HYPOTHESIS ON THE CONSTITUTION OF THE ZODIACAL LIGHT.

Many hypotheses have been made for the purpose of explaining the discrepancies shown in the preceding article. In general, either the assumption is made that NEWTON's Law of Gravitation is not strictly accurate* or else that certain additional matter in the solar system must be considered whose attraction has not hitherto been allowed for.† The most recent and the most plausible investigation of the second kind is that effected by SEELIGER[10], [11], [12] who seeks to account for all of the appreciable discrepancies by the perturbing effect of the cloud of particles known as the zodiacal light.

What the true form of this cloud is, and still more, what the law of the distribution of its density is, is very uncertain.‡ SEELIGER assumes that it can be roughly conceived as made up of two homogeneous ellipsoids of revolution whose semi axes have the values 0.24 and 1.20, respectively. Both the eccentricities of these ellipsoids and the position of the equator of the outer one can vary within wide limits without greatly altering the values of the perturbations which they produce; the distance from the focus to the center in each of them is arbitrarily chosen as equal in length to ten times the semi minor axis, and the equator of the outer one is assumed to be coincident with the plane of the equator of the sun. The respective densities and also the two constants which define the equatorial plane of the first ellipsoid remain as unknown quantities whose values are to be determined.

* See Tisserand's Mecanique Celeste, Vol. IV, Pages 494–542.
† See Newcomb's " Astronomical Constants. . . ."[17], Pages 110–120.
‡ See the article, ''The Zodiacal Light'' by Newcomb, in the Encyclopaedia Britannica, Vol. XXVIII.

From the known formulas which express the attraction exerted by an ellipsoid upon a point either wholly within or without its surface, the expression for the perturbing force in any case can readily be written, and from this the equations for the variations of the various elements are derived, each equation containing five unknown quantities whose values are to be so determined as to best account for the excess of the variations observed over those heretofore obtained from the theory. As the ellipsoids are assumed to be symmetrical with respect to their axes of rotation, however, they will cause no appreciable perturbation of any eccentricity. The variation of the obliquity of the Earth's orbit was also not considered by SEELIGER.

There remain therefore but ten discrepancies to be represented; namely, those of the four perihelia, those of the three nodes and those of the three inclinations. These ten discrepancies form the absolute terms of ten corresponding equations which contain five unknown quantities. It is to be noticed that in the "Astronomical Constants . . . " two tables of the theoretical variations are stated by NEWCOMB; the first, on page 109, are those computed from the values of the various masses assumed in Chapter V; the second, on page 185, are those computed from the definitively adopted masses. The latter values of the masses are in closer accordance with those assumed in the present paper than the former; the first values are, however, the ones adopted by SEELIGER in the computation.

The final results are as in the following table:

	Newcomb.	Method of Gauss.	Per. caused by Zod. L't.	Final Residuals.		Prob. Errors.
				Newcomb.	Meth. of Gauss.	
MERCURY.						
$ed\pi$	$+8.64''$	$+9.33''$	$+8.49''$	$+0.15''$	$+0.84''$	$\pm0.29''$
$\sin id\Omega$	$+0.61$	$+0.45$	$+0.62$	$+0.01$	-0.17	±0.54
di	$+0.38$	$+0.40$	$+0.49$	-0.11	-0.09	±0.35
VENUS.						
$ed\pi$	-0.10	-0.09	$+0.05$	-0.15	-0.14	±0.17
$\sin id\Omega$	$+0.60$	$+0.63$	$+0.60$	0.00	$+0.03$	±0.22
di	-0.05	-0.05	$+0.20$	-0.25	-0.25	±0.11
EARTH.						
$ed\pi$	$+0.23$	$+0.23$	$+0.09$	$+0.14$	$+0.14$	±0.09
MARS.						
$ed\pi$	$+0.76$	$+0.81$	$+0.56$	$+0.20$	$+0.25$	±0.24
$\sin id\Omega$	$+0.03$	-0.09	$+0.21$	-0.18	-0.30	±0.14
di	-0.01	$+0.03$	-0.01	0.00	$+0.04$	±0.15

The first two columns of the table contain the residuals from the masses employed in the present paper; the third column states the perturbations caused by the zodiacal light when its elements are derived from the residuals of NEWCOMB's first tabulation. As the five elements were so determined as to represent NEWCOMB's first residuals as accurately as possible, their agreement with these is naturally more exact than with the values here stated. Thus the first agreement for the motion of Mercury's perihelion is exact while here the discrepancy is considerable. On the other hand, the greatest discrepancy when the results are compared with the first tabulation, and which occurs in the motion of the node of Mars, is slightly lessened when the new masses are employed.

As the five elements were determined to represent the ten residuals of NEWCOMB's computation as accurately as possible, the numbers of the fourth column are, as might have been expected, generally smaller than those of the fifth. It may justly be inferred, however, that SEELIGER's hypothesis is capable of greatly reducing those discrepancies whose values are sufficiently large to establish their reality, without at the same time unduly increasing any of the smaller ones.

The last column contains NEWCOMB's estimate of the total probable errors arising both from the errors of observation and from the uncertainties in the values of the adopted masses.

The elements of the zodiacal light derived by SEELIGER are as follows:

Density of inner ellipsoid $= 2.52 \times 10^{-11}$ times the Sun's density.
Density of outer ellipsoid $= 0.0026 \times 10^{-11}$ times the Sun's density.
Total mass $\qquad = 35000 \times 10^{-11}$ times the Sun's mass.
Inclination of equator of I $= 6°.95$
Longitude of node of I $\quad = 40°.03$.

The unit of time throughout this article is the Julian Century.

BIBLIOGRAPHY.

LIST OF WORKS ON GAUSS'S METHOD AND RELATED SUBJECTS WHICH ARE REFERRED TO IN THE PRECEDING PAGES.

1. GAUSS. Determinatio Attractionis quam in punctum quodvis positionis datae exerceret planeta si ejus massa per totam orbitam ratione temporis quo singulae partes describuntur uniformiter esset dispertita. Werke, Vol. III, pages 333–357.
2. NICOLAI. Neue Berechnung der Secular Anderungen der Erdbahn. Bode's Astronomische Jahrbuch, 1820, pages 224–226.
3. CLAUSEN. Alia solutio problematis a celeberrimo Gauss in opera "Determinatio attractionis. . . " tractati. Crelle's Journal, Vol. VI, 1830, page 290.
4. —— Bestimmung der Bahn und der Umlaufszeit des Tuttle'schen Cometen. Beobachtungen der Kaiserlichen Universitats Sternwarte Dorpat, Vol. XVI.
5. BOUR, EDMOND. Thesis presentees a la faculte des sciences de Paris, 1855.

6. ADAMS. On the November meteors. Monthly Notices, Vol. XXVII; Collected Works, Vol. II, pages 194–200.

7. LEVERRIER. The secular perturbations of the elements of the orbits of the planets. Annales de l'Observatoire de Paris. Mercury, Vol. V, pages 6 and 7; Venus, Vol. VI, page 6; The Earth, Vol. IV, pages 11 and 12; Mars, Vol. VI, page 189.

8. HILL. On Gauss's method of computing secular perturbations. Astronomical Papers of the American Ephemeris, Vol. I, pages 317–361.

9. SEELIGER. Ueber das von Gauss herruhrende Theorem die Sacularstorungen betreffend. Astronomische Nachrichten, Vol. XCIV, 1879.

10. —— Ueber die sogenannte absolute Bewegung. Sitzungsberichte der konigliche Akademie der Wissenschaften zu Munchen. Vol. XXXVI, pages 85–137.

11. —— Ueber die empirischen Gleider in der Theorie der Bewegung der Planeten Merkur, Venus, Erde und Mars. Vierteljahrschrift der Astronomischen Gesellschaft, Vol. XLI, pages 234–240.

12. —— Das Zodiakallicht und die empirischen Gleider in der Bewegung der inneren Planeten. Sitzungsberichte der koniglichliche Akademie der Wissenschaften zu Munchen. Vol. XXXVI, pages 595–622.

13. CALLANDREAU. Calcul des variations seculaires des elements des orbites. Annales de l'Observatoire de Paris, 1885, Vol. XVIII.

14. TISSERAND. Traite de Mecanique Celeste. Vol. I, pages 431–442; Vol. IV, pages 494–542.

15. NEWCOMB. Secular variations of the orbits of the four inner planets. Astronomical Papers of the American Ephemeris, Vol. V, pages 301–378.

16. HILL. A new theory of Jupiter and Saturn. Astronomical papers of the American Ephemeris, Vol. IV

17. NEWCOMB. The elements of the four inner planets and the fundamental constants of astronomy. 1895. Supplement to the American Ephemeris and Nautical Almanac, 1897.

18. HALL, ASAPH, JR. Secular perturbations of the Earth from the action of Mars. Astronomical Journal, No 244.

19. SEE. Secular perturbations of Uranus from the action of Neptune. Astronomical Journal, No. 316.

20. INNES. Secular perturbations of the Earth from the action of Mars. Monthly Notices, Vol. LII, Nos. 2 and 7.

21. —— Secular perturbations of the Earth from the action of Venus. Monthly Notices, Vol. LIII, No. 6.

22. —— Tables to facilitate the application of Gauss's method. Monthly Notices, Vol. LIV, Nos. 5 and 6.

23. TURNER, ARTHUR B. Secular perturbations of Mars from the action of Jupiter. Thesis presented to the Faculty of Philosophy of the Graduate School of the University of Pennsylvania, 1902.

24. —— Secular perturbations of Mars from the action of Jupiter, computed by the method of Arndt. Astronomische Nachrichten, No. 3065.

25. BARTON, SAMUEL B. Secular perturbations of Mars from the action of Saturn. Thesis presented to the Faculty of Philosophy of the Graduate School of the University of Pennsylvania. 1906.

26. MERFIELD, C. J. The secular perturbations of Ceres from the action of Jupiter. Astronomische Nachrichten, No. 4215.

27. DZIEWULSKI, W. Sakulare Marstorungen des Eros. Cracovi, 1906.

28. HALPHEN. Traite des fonctions elliptiques. Part II, pages 310–328.

29. BRUNS. Ueber die Perioden der elliptischen Integrale erster und zweiter Gattung. Dorpat, 1875.

30. ARNDT, LOUIS. Recherches sur le calcul des forces perturbatrices dans la theorie des perturbations seculaires. Bulletin de la Societe des sciences naturelles de Neuchatel, Vol. XXIV, 1896.

31. INNES. The computation of secular perturbations. Monthly Notices, Vol. LXVII, pages 427–443.

32. ROBBINS, FRANK. Tables for the application of Mr. Innes' method. Monthly Notices, Vol. LXVII, pages 444–447.

33. MERFIELD, C. J. Extension of Mr. Robbins' tables to the value $i = 180°$. Monthly Notices, Vol. XLVIII, pages 605–608.

34. —— The secular perturbations of Eros. Astronomische Nachrichten, Nos. 4178–4179.

35. —— The secular perturbations of Iris. Astronomische Nachrichten, No. 4337.

36. —— The secular perturbations of Ceres. Monthly Notices, Vol. XLVII, pages 551–560.

37. HILL, G. W. The secular perturbations of the planets. American Journal of Mathematics, Vol. XXIII, page 317.

38. —— On the use of the sphero-conic in astronomy. Astronomical Journal, No. 511.

39. INNES. Jacobi's nome, q, in astronomical formulas, with numerical tables. Monthly Notices, Vol. LXII, pages 494–503.

www.ingramcontent.com/pod-product-compliance
Lightning Source LLC
Chambersburg PA
CBHW081337190326
41458CB00018B/6033